Illuminating Edison

The Genie of Menlo Park and the New York *Sun*, 1878-1880

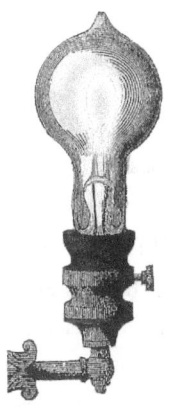

Edited with an Introduction by
Jerald T. Milanich

the PeppertreePress, LLC
Sarasota, Florida

For information regarding permission,
call 941-922-2662 or contact us at our website:
www.peppertreepublishing.com or write to:
the Peppertree Press, LLC.
Attention: Publisher
1269 First Street, Suite 7
Sarasota, Florida 34236

ISBN: 978-1-61493-727-2

Library of Congress Number: 2020913741

Printed August 2020

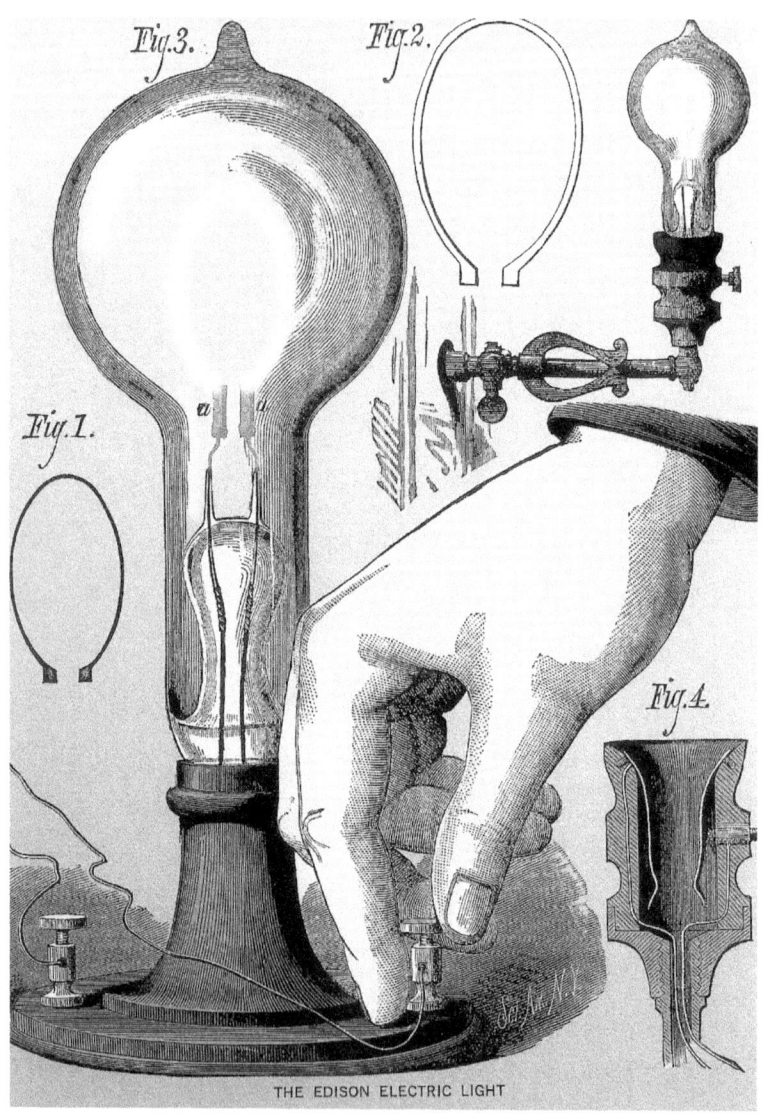

Fig.3. Fig.2.

Fig.1.

Fig.4.

THE EDISON ELECTRIC LIGHT

This drawing is from *Scientific American*, January 10, 1880, p. 19

CONTENTS

—INTRODUCTION—

The Genius, the Journalist, and the Sun

The close relationships among Thomas Edison, New York *Sun* editor Amos Cummings, and the *Sun* newspaper began with a hand-written note penned January 3, 1878 by Uriah H. Painter, correspondent for the Philadelphia *Enquirer* in Washington D.C. Painter, a friend of Edison's, was prompted by Amos Cummings to provide a letter of introduction to the New Jersey-based, soon-to-be famous inventor. Addressed to "Prof T A Edison, Menlo Park," Painter's handwritten note read:

> *Sir, Allow me to introduce Amos J Cummings of the New York Sun. He wants to see & and have a little talk with you over such matters as you can tell & show him that while they are familiar to you are yet untold wonders to the public.*

The next day Painter alerted Edison to Cummings's interest in a short note sent from Washington D. C.:

> *One of the NY Sun Editors wanted a letter of intro me to you & I gave him one. Thought you might want to have access to a live paper sometime & it would be well to know him.*

Edison, in a note back to Painter on January 7, assured Painter "[I] will treat *Sun* man well."

With the stage set, Amos Cummings wrote directly to "Prof. T. A. Edison" on January 8 enclosing Painter's letter of introduction. Cummings's note, on "Editor's Office of the *Sun*" letterhead, read:

Dear Sir,

I am a sort of a bluster-headed newspaper writer with letter to you from Mr. U. H. Painter of Washington, and want an interview for the columns of the Sun—*something about the secrets of electricity and so on. Will you please designate a time and place where I can spend an hour or more with you? Being acquainted with Stewart, Ass. Press Operator, I know some of your history and want to know you right bad.**

Yours Truly,
Amos J. Cummings

The introduction was made. Following Cummings's January 7 letter to Edison, the two exchanged more communications setting up their initial meeting, which took place February 16, 1878 in Menlo Park. At the time Cummings was 49 years of age and Edison had just turned 31.

Amos Cummings's first article highlighting Edison, "A Marvellous Discovery," was published in the *Sun* on February 22, 1878. The article went viral. It was picked up by other newspapers as well as by the *Scientific American Supplement*, a weekly periodical with a nation-wide distribution, and printed in the magazine's March 16, 1878 issue (with an attribution to the "N. Y. *Sun*"). Cummings and people across America were hooked on Edison.

In the article Cummings relates the story of Edison's phonograph recording "Mary Had a Little Lamb," an incident I recall reading about in a biography of Edison when I was in elementary school many decades after it took place!

Beginning in 1878 and continuing through 1880 (and beyond), Amos Jay Cummings, well-known journalist and

---

* George S. "Fattie" Stewart was a telegraph operator for Western Union whose offices were in New York City at 45 Broadway (see Chapter 2 below). The notes from Painter, Edison, and Cummings are archived in the Edison Papers at Rutgers University, https://edison.rutgers.edu/digital/.

senior editor for the *Sun*, and Thomas Alva Edison, who soon would become internationally known for his invention of the phonograph and his successful experiments with the electric light bulb and electricity, would meet on multiple occasions in Edison's Menlo Park laboratory. They also met in New York City. Those meetings provided the basis for at least fifteen articles written by Cummings and published in the pages of the *Sun* newspaper from 1878 through 1880 (a sixteenth and seventeenth were published in early 1881).

During the period 1878-1880, at least thirty-eight other articles about Edison and his inventions (some quite informative, others were short, or they were editorials) appeared in the *Sun*. Cummings certainly wrote or edited some of those articles. The *Sun* also ran a number of letters to the editor about Edison in those same three years.

Cummings's interest in Edison likely was piqued by two earlier *Sun* articles written about Edison's "speaking phonograph" (a recording device). The first was published in the *Sun* on November 6, 1877 ("Echoes from Dead Voices") and included a letter from Edward H. Johnson about Edison's phonograph; the letter would appear November 17 in *Scientific American*. Johnson was himself an inventor and an associate of Edison (Johnson is credited with inventing electric Christmas tree lights). The subheading for "Echoes from Dead Voices" read "Wonderful Possibilities of Mr. Edison's Latest Invention."

The second article in the *Sun* on January 2, 1878 ("A Marvellous Invention") also focused on Edison's phonograph, extoling its possibilities and recounting the results of a demonstration held at the Western Union Telegraph office in New York City. The sub-heading was "Remarkable Experiments with Mr. Edison's Phonograph—Astonishing Results."

Cummings and the *Sun*, more than any other journalist and newspaper, became Edison's public voice during the period 1878-1880 when Edison was perfecting his recording device and the electric light bulb, as well as generating and harnessing electricity for general use. Cummings's interviews with Edison and the resulting *Sun* articles were picked up across the country by other papers and received wide coverage. It was Cummings's articles that christened Edison the "Genie of Menlo Park," the "Napoleon of Science," and the "Inventor of the Age."

At a time when other scientists were labeling electricity a fraud and gas company executives were giving interviews to convince the public that the electric light could never replace their product, Amos Cummings's press coverage helped catapult Edison to the status of a modern rock star. At one time the public was provided with nearly hour-by-hour coverage chronicling Edison's laboratory trials of his light bulb. Could Edison perfect a bulb that could be manufactured in quantity, burn for long periods, and operate on relatively small amounts of inexpensive electricity sent door to door by wires? He could, and he did, much to the delight of Cummings and his readers.

Cummings's and the New York *Sun*'s journalistic relationship with "friend Tom" would provide Edison, his inventions, and his attempts to successfully market them with the sort of in-depth, positive press coverage that modern scientists can only dream about. Edison was well aware that stories in the popular media would draw attention from entrepreneurs and investors whom he needed for financial support. In Cummings he found the perfect media personality, a respected reporter who could write and who was well-known in journalistic, social, and political circles in New York City.

Cummings, credited by many as originating newspaper human-interest stories (one colleague said he invented "man bites dog" journalism), was almost as much a genius in his chosen profession as

Edison was in his. Certainly, Edison, who was brilliant, would have been successful in his endeavors without Cummings's intervention, but the stories Cummings wrote about him could only have added to Edison's public persona as a genius and inventor.

Who was Amos Jay Cummings and why did he choose to focus his considerable journalistic skill on Thomas Alva Edison? The answer to the latter question is easy: Cummings knew a good story when he saw one. Explaining who Amos Cummings was and how I learned of him requires a longer response.

I owe my discovery of Amos Cummings and the New York *Sun* articles about Edison to James G. Cusick, Collection Curator of the Library of Florida History at the University of Florida. Jim, knowing of my interest in Florida archaeology, had sent me a newspaper clipping from the New York *Weekly Sun*, a sister publication of the *Sun*. The nearly 12,000-word article, "Florida's Indian Mounds, the Mysterious Monuments of an Extinct Race," was published April 23, 1873. The article provided a wonderful description of Turtle Mound, a famed Indian shell mound in east Florida, along with mention of other sites. I set out to find who wrote it. Might there be other archaeological articles by the same author?

My two-week search took me on an incredible journey through library and on-line sources. By luck I came across an article quoted in a book that contained some of the exact same wording as in the *Weekly Sun* piece. That article was written by Ziska, an obvious penname. John Edward Haynes's slender 1882 volume *Pseudonyms of Authors: Including Anonyms and Initialisms* told me Ziska was a pseudonym used by Amos Jay Cummings, a New York *Sun* journalist. I had my author!

From an assortment of biographical dictionaries and similar books, I would learn that Cummings was a respected journalist, a hero of the Civil War who received the Medal of Honor, and a member of the United States House of Representative (first elected in 1886). In 1896 he was considered as a Vice Presidential running mate for William Jennings Bryant (he was not selected; Bryant lost to William McKinley in the November election).*

My attempts to learn more about Cummings took me to the on-line catalogue of the New York Public Library. There, much to my amazement, were listed four scrapbooks that had belonged to Amos Cummings. On a subsequent visit to the library to look at the scrapbooks I stopped off at the bound copies of the old card catalogue shelved just outside the elegant Rose Reading Room. Bingo! Eleven additional scrapbooks were listed under Cummings's name. Over a number of visits, I read all fifteen.

How did the fifteen scrapbooks end up in the New York Library? It is likely that after Cummings's death in 1902 they were kept by his widow Frances Caroline (nee Roberts) who continued to live in their town house on Charlton Street in New York City until she moved to Connecticut in 1910 or 1911 due to poor health. The scrapbooks must have been left in the house when she left (stored in the attic?). Years later the scrapbooks found their way to the Bargain Book Shop in Manhattan. According to notes in the New York Public Library catalogue, the scrapbooks were purchased in 1927 by a library employee and added to the library's collections.

The scrapbooks are a goldmine. They contain a large number of articles written by Cummings for the New York *Sun*, the *Weekly Sun*, and *Semi-Weekly Sun* (both of the latter reprinted articles from the *Sun*). Almost none of the many articles in the scrapbooks are

---

* The story of my search for Amos Jay Cummings can be found in Jerald T. Milanich, "The Historian's Craft," *Florida Historical Quarterly* 80(3):375-378; for additional biographical information about Cummings, see *A Remarkable Curiosity: Dispatches from a New York City Journalist's 1873 Railroad Trip across the American West* (Boulder: University Press of Colorado, 2008), and *The Florida Adventures of Amos Jay Cummings, 1873-1893* (Cocoa, FL: Florida Historical Society Press, 2016).

labeled with a publication date. The scrapbooks also have clippings from other newspapers and magazines which Cummings must have found interesting (did he write some of these other articles?), along with hundreds of articles from various United States newspapers that mention U.S. Representative Cummings. Most likely his congressional staff subscribed to a clipping service and compiled some of the later scrapbooks (those scrapbooks are arranged much more neatly than the others). The contents of the fifteen scrapbooks span more than a thirty-year period, from the early 1870s until just after Cummings's death. The last scrapbook contains obituaries and information about Cummings's funerals.

I learned of Cummings's Edison articles from scrapbook 4 (of the four listed on-line). Pasted on the inside cover is a 3½-by-7½-inch torn piece of paper with a hand-written, abbreviated list of some of the scrapbook's contents. Number 1 on the list is "Edison's Electric Light." Elsewhere in the scrapbook I found a list of seventeen articles about Edison, which are those referred to in the contents as "Edison's Electric Light." I am certain Cummings wrote all of them. None, except the very first one, are dated, nor, I would learn, is the list in chronological order. The first of the seventeen has the number 1 and the date February 22, 1878 hand-written on the top. The next fourteen, numbered 2-15 are all from 1878 and 1879. The last two, both unnumbered, are from 1881. I found clippings of the seventeen articles themselves in the scrapbook. Because of the way the articles were cut out of the newspaper, none have dates on them, nor is there any indication that they were published in the *Sun*, but I suspected strongly that they were.

At the time I was doing my research, the New York *Sun* newspaper had never been indexed or placed on-line. The New York Public Library's copies of the newspaper existed only on well-worn microfilm. Thankfully, I eventually discovered that the

Bobst Library at New York University had copies of the same reels of microfilm, and they were in better shape. Even so, reading (and copying) the microfilm was not an easy task, though the modern microfilm reader machines in the Bobst Library that allowed me to download parts of newspaper columns onto a thumb drive were a godsend. I spent weeks and weeks poring over microfilm, searching for Cummings's seventeen Edison articles that were in the scrapbook. I found all of them, along with a number of other *Sun* articles about Edison. It was quite an adventure, one that led to this book. [Since my original research, the *Sun* has been digitized and, in February 2016, placed on-line, thanks to the Northern New York Library Network and the New York Public Library. The newspaper now can be searched at the NYS Historical Newspapers website: https://nyshistoricnewspapers.org/.]

After their initial meeting in early 1878 Thomas Edison and Amos Cummings remained friends for more than two decades. Other journalists soon realized that Cummings and the *Sun* had a special "in" with Edison. On more than one occasion journalists asked Cummings's and the *Sun*'s help in gaining access to Edison.

Over the many years of their friendship, Edison and Cummings exchanged notes on multiple occasions, at times sending communications by telegraph. A note could be carried by an office boy from the *Sun* offices to Western Union, six blocks away in lower Manhattan, where it was sent over the wires to Menlo Park (and vice versa). It was not as fast as a modern cellular phone, but still a message could go back and forth in a matter of minutes.

Some correspondence between the two men was business-oriented and some social. They clearly liked one another. Here are transcriptions of parts of a few of their notes, taken from items in the Edison Archive at Rutgers University.

In this letter penned by Cummings on "Editor's Office of The *Sun*" letterhead and datelined New York, December 29, 1882, Cummings warns Edison about a representative of William E. Sawyer's who had come to the *Sun* offices. Sawyer was one of Edison's chief American rivals with whom he was involved in litigation. "Uncle" may be a euphemism for attorney:

My Dear Tom:

I write you this because it may serve to put you on your guard. Sawyer's uncle has been to the office inquiring for the man who wrote the article in the Sun on Nov. 25, '78, headed "The New Electric Lights;" also the article on April 12, 1879, headed "Edison's Electric Light;" also the one on Nov. 15, 1878, headed "The New Electric Light." He said that he wanted the man to verify the reports in establishing a patent. Of course, "the man" declined....

Yours, always,
Amos

The "man," of course, was Amos Cummings.

A request of some sort from Edison to Cummings elicited this response (again, on *Sun* letterhead) date June 22, 1883:

Dear Tom:

All right. I'll do what I can, and as soon as possible. If you were an Irishman, however, God would have been good to you.

Truly,
Amos

In 1885 Cummings consulted with Edison about a planned demonstration of his electric light in New York City. At one point, Cummings was the president of the New York Press Club, which is mentioned in this January 4 letter; I do not know what U. S. is in reference to, though it may have been the *Sun*'s offices:

Dear Tom:

I think it would be a good thing to have the Edison Electric Light in the Press Club instead of the U. S. They say the Club owns the wires and fixtures. The bulbs are like yours. I wish you would come down some day, and take a look at it—or some night. Set your own time & I'll be with you;

> *Yours truly,*
> *Amos J. Cummings*

P.S. And while you're about it you might become a member of the Club. $10 initiation & $12 a year dues—won't break you.

Edison responded on January 12:

Amos J. Cummings, Esq.
My Dear Cummings

I was out of town when your note arrived. I will meet you tomorrow Tuesday night about 7:30 & pay a visit to the Press Club with you. Where shall I meet you? I shall be glad to become a member of the Club.

> *Yours truly,*
> *Thos. A. Edison*

On October 29, 1886 when he was a candidate for the U.S. House of Representatives, Cummings sent a request for a campaign contribution to Edison. Upon receiving the letter Edison wrote $100 across the top ($100 in 1886 is worth about $2725 today). The letter read:

Personal
Dear Tom:

I'm running for Congress, as you probably have seen. The assessments are very heavy, and I have little money. If you can aid me a little I would be grateful.

Yours, truly,
Amos J. Cummings

Four years later, in response to something (a request?) Cummings sent to Edison, Edison typed this answer (June 20, 1890). At the time Cummings was in his second term in the U.S. Congress.

My dear Cummings,

This is O. K. Send in your stock. I quit drinking beer—only take whisky now.

Yours very truly,
Thomas A. Edison

Despite being friends with Edison, Amos Cummings apparently was not above making money off the inventor's genius, as suggested by a brief November 4, 1886 article in the Washington *Post*. It likely was written by Cummings himself. It reads like his prose and he occasionally wrote for the *Post*. I also know that he oversaw the placement of similar, tongue-in-cheek articles in local newspapers along the way when he traveled across the United States from New York to California in 1873. Those articles generally extolled his skills as a reporter (and a fisherman). The 1886 article reads in part:

AMOS CUMMINGS, the new Democratic Representative in Congress from the New York *Sun*, thoroughly deserves his honors. He is one of the most skillful and accomplished journalists in New York.... Six years ago he made a lot of money in the first boom of Edison's electric-light stock, when it rose in a month from $80 a share to $3000.... He is a picturesque and humorous writer, and has always been widely quoted.

It should be noted that despite being friends, there must have been a certain tension between Thomas Edison and Amos Cummings. The reporter wanted all the information he could get from his visits with Edison to sate the public's thirst for news about Edison and his latest inventions. Edison, who seemingly loved talking to Cummings about his latest endeavors, had to be careful not to reveal too much. Cummings's articles traveled far beyond New York City, including to France and England, where inventors read and used the information to advance their own interests. Some stole his ideas and filed patents under their own name. As Cummings notes in Chapter 6 below, people were not shy in trying to steal Edison's "thunder."

Cummings's death in 1902 in a Baltimore hospital was national news. The New York *Times* and the Washington *Post*, as well as the *Sun*, carried multiple stories about his life. He was honored with a state funeral in the Hall of Representatives in the Capitol in Washington, followed by multiple funerals in New York City and another at grave side at Clinton Cemetery in Irvington, New Jersey, the town where he had once lived when he was a schoolboy. According to a date on his tombstone Cummings was born in 1838. Edison would outlive Amos Cummings by 29½ years. At the time he died Edison held nearly 1,100 patents in the United States and more than 1,200 in thirty-four other countries around the globe.

As noted above, Amos Cummings wrote at least seventeen articles about Edison that were published in the New York *Sun* from 1878 into early 1881; additionally, thirty-eight other articles about Edison appeared in the *Sun* from 1878-1880. Selecting which of these *Sun* articles to reproduce in this book was not an easy task. I knew I wanted twelve of the seventeen articles that I was certain Cummings had written (two were not that interesting; number 15, dated July 11, 1879, was too short for a book chapter, and, as stated above, the sixteenth and seventeen items in the list were from 1881, and I wished to focus only on the years 1878-1800). Because I could not be sure which of the other thirty-eight *Sun* articles (1878-1880) might be attributed to Cummings, I simply chose twelve I liked best.

The twelve articles from Amos Cummings's scrapbook list and the other twelve total twenty-four, each of which is presented here as a separate chapter, organized chronologically. The last article, Chapter 24, I selected both because I enjoyed it and because it adds a little more star power to the book. You will see why when you read it.

The articles offer extraordinary, almost firsthand accounts of Edison's inventive genius and they catch the public excitement surrounding his recording device, the invention of a utilitarian electric light bulb, and the fine tuning of electricity as a power source for everyday life. The *Sun* articles are not only about Edison's successes; they tell us about international intellectual espionage: efforts by others to steal or lay claim to Edison's ideas and inventions. Edison was continually forced to immediately file patents on his inventions in both America and Europe before others stole his ideas.

One thing I learned from these twenty-four *Sun* articles is that Edison's famed laboratory in Menlo Park was nothing like the high school and college labs where I studied chemistry and physics. Menlo Park was an industrial-sized operation with multiple buildings, quantities of equipment, and numerous assistants. Visiting scientists and inventors regularly passed through. The physical plant and the research were fueled by large amounts of money gleaned from Edison's patents, his commercialization of inventions, and investors. It was a complex set-up and Cummings's descriptions provide a "you are there" experience.

Cummings also interviewed associates of Edison, some from his days before Menlo Park, to create a written portrait of the inventor, one that includes anecdotes like Edison's invention of the first electric "bug zapper." These and the other *Sun* descriptions of Edison's persona are simply charming. The reading public must have been enthralled with the *Sun*'s treatment of Edison, especially as they came to realize the changes the inventor was bringing into their lives.

I hope that you will find the articles by Amos Cummings and other New York *Sun* reporters as enjoyable as readers of that newspaper did nearly a century and a half ago. After reading these stories you likely will never look at an electric light bulb without thinking of Thomas Edison and his experiments at Menlo Park.

ACKNOWLEDGMENTS

My original plan was to optically scan the articles from the New York *Sun* and turn them into word processing files; it failed miserably. The images derived from the microfilm were, by and large, horrible. Enter Andria Kuzeff who deciphered the images and transcribed them into Microsoft Word files. She deserves my deepest thanks. I've often wondered what Thomas Edison would have thought of her computer and word-processing software.

I am also grateful to the staffs of the New York Public Library's Microforms Room and the Bobst Library's Microforms Center (at New York University). Patrick Payne, a graphics genius, once again made old images look new. He has my gratitude and admiration.

This is my second book-publishing project with Peppertree Press; they have been terrific. Thank you!

Thomas Alva Edison sitting on top of the world, wired to the cosmos through his telephone, phonograph and "talking machine." The caricature by J.C. Fireman first appeared ca. 1900 in the *Mail and Express*, a New York newspaper formerly (and later) named the *Evening Mail*. The drawing was published about 1905 in *Men of Affairs: The Evening Mail*, an undated book containing caricatures by Fireman and two other artists.

Amos Jay Cummings

Amos Jay Cummings portrayed in the *Sun,* March 25, 1888 (p.8). The drawing, headlined "HERE HE IS! The Journalist Statesman," accompanied a letter "TO THE EDITOR OF THE SUN" that sang the praises of Cummings and requested "a good likeness of this eminent pen artist." Cummings was known to use pseudonyms to write letters to the *Sun* editor (himself).

The Sun.

1. A MARVELLOUS DISCOVERY.*

THE BEDLOE'S ISLAND STATUE TO TALK AND WHISTLE.

A MAN OF THIRTY-ONE REVOLUTIONIZING THE WHOLE WORLD—THE UNTOLD WONDERS OF THE SPEAKING PHONOGRAPH—A VISIT TO PROF. THOMAS A. EDISON OF MENLO PARK, N. J.

The writer visited Menlo Park, N. J., on Saturday to chat with Prof. Thomas A. Edison. This gentleman is the inventor of the automatic telegraph, quadruplex and sextuplex despatches, the carbon telephone, the stock indicator, the electric pen, the airophone, the marvellous speaking phonograph, and a score or more of similar machines. He is also the discoverer of the electromotograph, by which despatches may be telegraphed without magnetism. Scientific men regard it as his greatest discovery, and predict that it will some day prove of immense value.

Menlo Park is a small place on the line of the New York and Philadelphia Railroad, two miles north of Metuchin [sic]. Mr. Edison's manufactory stands forty rods west of the depot. A high

---

* From the New York *Sun*, Friday February 22, 1878, p. 3; reprinted in *Scientific American Supplement* March 16, 1878, pp. 1828-29; clipping in Cummings scrapbook 4, New York Public Library (NYPL). In editing these articles from the *Sun* I have corrected obvious typos but retained original spellings; e.g., I changed "hte" to "the," but I left other spellings as Cummings wrote them; e.g., "marvelous" appears as "marvellous," "center" as "centre," "sulphur" as "sulfur," "everyone" as "every one," "today" as "to-day," and "Metuchen" (New Jersey) is "Metuchin," etc. "Percent" is printed as "per cent.," an abbreviated form of "per centage." I removed the period after "cent" to make reading easier. In some articles the same word is spelled two different ways.

bank shuts out the view from the car windows. The building is a long wooden structure, something like an old-fashioned Baptist tabernacle. It faces to the east. Nine lightning rods pierce the sky above it. A dozen telegraph wires are led into it by sentry-like poles connecting with the main line along the railroad. The front doors open directly into the office. The writer entered. A man sat at a table studying a mechanical drawing. An inquiry for Mr. Edison drew from him the words, "Go right upstairs, and you'll find him singing into some instrument."

The stairs were climbed, and the writer stepped into a long room forming the second story. It was an immense laboratory, filled with electrical instruments. A thousand jars of chemicals were ranged against the walls. A circle of kerosene lamps was smoking viciously on an empty brick forge. Their chimneys were the essence of blackness. There was no disagreeable smell, for the smoke was borne off by the draft of the forge. An open rack loaded with jars of vitriol stood in the middle of the room, and the rays of the sun struck through them, flecking the floor with green patches. The western end of the apartment was occupied by telephones and other instruments, and there was a small organ in the southwestern corner.

Prof. Edison was seated at a table near the centre of the room. He looked like anything but a professor, and reminded me of a boy apprentice to an iron moulder. His hands were grimy with soot and oil; his straight dark hair stood nine ways for Sunday; his face was entirely beardless, but sadly needed shaving; his black clothes were seedy, his shirt dirty and collarless, and his shoes ridged with red New Jersey mud; but the fire of genius shone in his keen gray eyes, and the clean cut nostrils and broad forehead indicated strong mental activity. He seems to be always looking for something of great value, and to be just on the point of finding it. Unfortunately he is quite deaf, but this infirmity seems to increase his affability and playful boyishness. A man of common sense would feel at home with him in a minute; but a nob or prig would be sadly out of place. Though but 31 years old, the occasional gleam of a silvery hair tells the story of his application.

The Professor was manipulating a machine upon the table before him. He had something resembling a gutta-percha mouthpiece of a speaking-tube shoved against a cylinder wrapped in tinfoil, which he turned with a crank. The small end of a tin funnel was clapped over the mouthpiece, and strange ventriloquial words were issuing forth from it. He shook hands, and pointing to the instrument said: "This is my speaking phonograph. Did you ever see it and hear it talk?"

The reply was a negative. Thereupon, he picked up the gutta-percha mouthpiece, saying, "This mouthpiece is simply an artificial diaphragm. Turn it over," suiting the action to the word, "and you see this thin disk of metal at the bottom. Whenever you speak in the mouthpiece the vibrations of your voice jar this disk, which, as you see, has in its centre a fine steel point. Now for the other part of the machine. Here is a brass cylinder grooved something like the spiral part of a screw, only much finer. I wrap a sheet of tinfoil around the cylinder, and shove the mouthpiece up to it so that the tiny steel point touches the tinfoil above one of the grooves. I then turn the cylinder with a crank, and talk into the mouthpiece. The vibrations arouse the disk, and the steel point pricks the tinfoil, leaving perforations resembling the old Morse telegraphic alphabet. They are really stereoscopic views of the human voice, recording all that is said, with time and intonations. It is a matrix of the words and voice, and can be used until worn out. Now let us reset the cylinder, so that the steel point may run over the holes or alphabet made when we talked in the mouthpiece. The thin metal disk rises, and, as the steel point trips from perforation to perforation, opening the valves of the diaphragm, the words, intonation, and accent are reproduced exactly as spoken. For instance, before you came up, I was talking to the instrument, and here is the matrix or stereoscopic view, if you please, of what I said," putting his finger on the tinfoil which still remained on the cylinder. "Now I reset the instrument," sliding the cylinder to the right. "Here the steel point starts at the same spot as when I talked through the mouthpiece, but its action is now controlled by the perforated alphabet. It repeats what I said. I use this sort of an ear trumpet to bring out the sound, so that you can hear it more distinctly. Listen."

He placed the small end of the funnel over the mouthpiece, shoved the mouthpiece against the cylinder, and turned the crank. The following words chased each other out of the funnel:

Mary had a little lamb,
Its fleece was white as snow,
And everywhere that Mary went
The lamb was sure to go—to go—to go—
Ooh ooh ooh—ah!
Cockadoodle doo—ah!
Tuck—a—tuck—a—tuck!
Tuck—ah! tuck—ah!

The cylinder was again set back, and the crank turned very slow. The effect was ludicrous, for the Professor had originally pronounced the words with great gravity and dignity, and the drawling way in which the instrument repeated them would have made a horse laugh. The cylinder was then turned very fast, and the words flew out of the funnel so fast that they struck the ear in a confused mass. But a most extraordinary effect was produced when the Professor turned the cylinder backward. It said:

Go to sure was lamb the,
Went Mary that everywhere and,

> *Snow as white was fleece its,*
> *Lamb little a had Mary.*

All this with profound gravity, as if the fate of the world depended upon the accent and pronunciation. Mr. Edison then tore off the tinfoil and wrapped a fresh sheet around the cylinder. One of old Mother Goose's rhymes was murmured into the mouthpiece, and its alphabet pricked out by the action of the steel point. The cylinder was then reset, and the crank turned, with the following result:

> *Rub a dub dub,*
> *Three men in a tub,*
> *And who do you think was there?*
> *The butcher, the baker,*
> *The candlestick maker,*
> *They all jumped out of a rotten potato.*

The instrument is so simple in its construction, and its workings so easily understood, that one wonders why it was never before discovered. There is no electricity about it. It can be carried around under a man's arm, and its machinery is not a fiftieth part as intricate as that of a sewing machine. It records all sounds and noises. The Professor blew in it at intervals, and the matrix recorded the sound and returned it. He whistled an air from the "Grande Duchesse," and back it came clear as a fife, and in perfect time. He rang a small bell in the funnel. The vibrations were recorded, and on resetting the cylinder, the tintinnabulatory sounds poured out soft and mellow. Mr. Edison coughed, sneezed, and laughed at the mouthpiece, and the matrixes returned the noises true as a die. But, most remarkable, the instrument sent back the voices of two men at the same time. To illustrate: The Professor, in a deep voice, recited in the mouthpiece the first verse of "Bingen on the Rhine." A matrix was obtained, the machine reset, the funnel placed in position, and the crank turned. The words came out as though some tragedian was endeavoring to affect an audience to tears:

A soldier of the legion lay dying in Algiers,
There was lack of woman's nursing,
there was lack of woman's tears,
But a comrade stood beside him while his life blood ebbed away,
And bent with pitying glances to hear what he might say.
The dying soldier faltered, and he took that comrade's hand,
And he said, "I never more shall see my own, my native land;
Take a message and a token to some distant friends of mine,
For I was born at Bingen—at Bingen on the Rhine.

While these affecting words were pouring out, the Professor shouted into the funnel several petulant exclamations. At the close of the verse the cylinder and its matrix were reset, and the recitation again came out of the funnel, interruptions and all, as follows:

A soldier of the legion lay dying in Algiers,
—"Oh shut up!"— —"Oh, bag your head!"
There was lack of woman's nursing, there was lack of
— — — "Oh, give us a rest!" — — — — woman's tears
— — "Dry up!"
But a comrade stood beside him while his life blood ebbed — — —
"Oh, what are you giving us!" — — —
"Oh, cheese
away.
it!"
The dying soldier faltered, and he took that comrade's
— — — — — "Police! Police!" — — — — — — "Po-
hand,
lice!"
And he said, "I never more shall see my own, my native

— — — *"Oh, put him out!"* — — — — — *"Oh, cork your-*
land."
self!"

It is impossible to describe the ludicrousness of the effect. The Professor himself laughed like a boy. One of his assistants told a story concerning a trap laid for a well-known divine, who was skeptical regarding the capabilities of the instrument, and evidently had a suspicion that the Professor was a ventriloquist. He wanted to talk into the mouthpiece himself, and see if his own words would be recorded and repeated. A matrix was put on the cylinder that had been used once before. The Doctor repeated a Scripture quotation, and, to his great astonishment, it came out as follows:

> *He that cometh from above is above all ["Who are you?"];*
> *he that is of the earth ["Oh, you can't preach!"] is earthly and*
> *speaketh of the ["I think you're a fraud!"] earth; he that cometh*
> *from heaven is above all. And what he has seen and heard*
> *["Louder, old pudding head!"] that he testifieth; and no man*
> *receiveth his testimony ["Oh, go and see Beecher!"].*

The possibilities and capabilities of this remarkable instrument are wonderful. Dolls and toy dogs can be made to recite nursery ballads, and wax figures of notables can use the voice and language of their originals. A prominent showman has already taken steps toward the formation of a museum of wax figures similar to Madame Tussaud's in London. All the figures are to speak. Matrixes of the voice and words of a gentleman whose imitations of Edwin Forrest are astonishing are to be secured and placed in the breast of a wax statue of the great tragedian. The voice and outward appearance of Mr. Forrest are to be perfectly copied.

"Why," says Mr. Edison, "Adelina Patti can sing her sweetest arias, and by this instrument we can catch and reproduce them exactly as sung. The matrixes can be copied the same as stereoscopic views, and millions sold to those owning the machine. A man can sit down in his parlor at night, start his phonograph, and enjoy Patti's singing all the evening if he chooses. The same with Levy's cornet playing. A matrix of his solos can be produced, and a million copies taken, and Levy's solos and Patti's arias can be given ten thousand years from now as perfectly and accurately as when these great artists were alive. If the last benediction of Pope Pius had been taken by the phonograph, the matrix could have been duplicated, and every true Roman Catholic on the face of the earth might have heard the benediction in the Pope's own voice and accentuation. There was a fortune in it. The matrixes could have been sold at five dollars apiece.

"Poor churches in the country," continued the Professor, "might have these machines rigged up over their pulpits, and by using the proper matrixes, could have Dr. Chapin, Dr. Bellows, Beecher, or any other great theological light expound to them in their own voices every Sunday. Thus the poor churches would save their money, and get rid of their poor preachers. Nor is this all. A man in Europe has invented a machine by which he takes an instantaneous photograph. Let us suppose that he photographs Dr. Chapin every second, and we take down his sermon on the matrix of the phonograph. The pictures and gestures of the orator, as well as his voice, could be exactly reproduced, and the eyes and ears of the audience charmed by the voice and manner of the speaker.

"Whole dramas and operas," continued Mr. Edison, his eyes sparkling with excitement, "can be produced in private parlors. The instrument can be used in a thousand ways. Say I hire a good elocutionist to read David Copperfield or any other work. His words are taken down by machine, and thousands of matrixes of David Copperfield produced. A man can place them in the machine, and lie in bed, while the novel is read to him by the instrument with the finest grade of feeling and accent. He can make it read slow or fast, can stop it when he pleases, and go back and begin again at any chapter he may choose. I could fix a machine in a wall, and by resonations any conversation in a room could be recorded. Political secrets and the machinations of Wall street pools might be brought to light, and the account charged to the devil. Kind parents could lie in bed and hear all the spooney courtship of their daughters and lovers. A man who loved the music of the banjo or the fiddle could buy his matrix and listen to Horace Weston or Mollenhauer whenever he liked. He could have the whole of Theodore Thomas's orchestra if he wanted it.

"To a certain degree," said Mr. Edison, "the speaking phonograph would do away with phonography. A man could dictate to his machine whenever he pleased, turn the machine over to an amanuensis, and let him write it out. A lawyer through he machine might make an argument before a court, even if he had been in his grave a year. An editor or reporter might dictate a column at midnight and send the machine up to the compositor, who could set the type at the dictation of the machine without a scrap of manuscript. I tell you there is no limit to the possibilities of the instrument."

At this point in the conversation the Professor sat down at his table and hallooed "Mad dog!" "Mad dog!" "Mad dog!" into the phonograph a half dozen times, and then amused himself by turning the crank backward. Then he made the instrument tell the old affecting story of Archibaldas Holden, and lay back and laughed heartily. We asked how soon the phonograph would be thrown upon the market.

"We expect to offer them for sale within two months," said the Professor. "The price of the finest machine will be about $100, but we shall sell inferior ones at a much lower price. The matrixes will be for sale like sheets of music, and can be used upon all the machines."

One of the remarkable features of the invention is the fact that the diaphragm can be placed in steam whistles and made to talk like a calliope. The captains of ships at sea miles away from each other could converse without trouble and correct their chronometers. The steam whistles would throw any voice into articulated speech. With a metal diaphragm in the whistle of a locomotive the engineer could roar out the name of the next station in a voice so loud that it could be heard by every passenger on the train and by every man within a distance of two miles. Placed in a steam fire engine, the chief engineer could talk to every foreman in the department without difficulty, no matter how great the uproar. A machine might be put up in the Jersey City Railroad depot that would shout "This side for Newark, Elizabeth, Rahway and New Brunswick! Train on the left for Philadelphia, Baltimore and Washington! Show your tickets!"

"Why," said the Professor, "I could put a metal diaphragm in the mouth of the Goddess of Liberty that the Frenchmen are going to put up on Bedloe's Island that would make her talk so loud that she could be heard by every soul on Manhattan Island. I could drop one in a calliope and set it talking so that men could hear it miles away. Within two years you will find the machine used for advertising purposes. It will be sitting in the windows of stores on Broadway and other streets singing out, 'Babbitt's best soap,' 'New York SUN—price two cents,' 'Brandreth's Pills,' 'Longfellow's Poems,' 'Ten cents for a shave!' and so on. There is no end to its uses. It will sing songs and whistle. A man has already made application to use the phonograph in cabs, so as to record the complaints of passengers. The Ansonia Clock Company of Connecticut have one in their manufactory this minute, and it shouts 'Twelve o'clock' and

'One o'clock!' so loud that it is heard two blocks off. One might be used as an alarm clock. If its owner wanted to get up at a certain time in the morning, he could set the alarm, and at the appointed hour the machine would scream, 'Halloo, there! Five o'clock! What's the matter with you? Why don't you get up?'"

The Professor calls the machine applied to steam whistles the airophone. He is now constructing one to put up in front of his manufactory, and intends to make it talk so that it can be heard two miles. He says "Old Bill Allen of Ohio will be nowhere."

Several of his speaking phonographs have been sent to England, where they have created a profound sensation. Mr. Edison says that he received a cable dispatch on Friday last, offering him £3,000 and half the profits for the right to sell the instrument in that country.

"How did you discover the principle?" asked the writer.

"By the merest accident," said the Professor. "I was singing to the mouthpiece of a telephone, when the vibrations of the voice sent the fine steel point into my finger. That set me to thinking. If I could record the actions of the point, and send the point over the same surface afterward, I saw no reason why the thing would not talk. I tried the experiment first on a strip of telegraph paper, and found that the point made an alphabet. I shouted the words 'Halloo! halloo!' into the mouthpiece, ran the paper back over the steel point, and heard a faint 'Halloo! halloo!' in return. I determined to make a machine that would work accurately, and gave my assistants instructions, telling them what I had discovered. They laughed at me. I bet fifteen cigars with Adams here [Adams was lying on the table listening to the conversation—REP.] that the thing would work the first time without a break, and won them. I bet two dollars with the man who made the machine, and won them also. That's the whole story. The discovery came through a pricking of the finger."

Here Mr. Edison, in a deep bass tone, shouted in the instrument:

"Nineteen years in the Bastile!
I scratched a name upon the wall,
And that name was Robert Landry,
Parlez vous Français? Si habla Español.
Sprechen sie Deutsch?"

And the words were repeated, followed by the air of "Old Uncle Ned," which the Professor had sung.

On being questioned concerning his telephone, the Professor said: "I went to work before Prof. Bell. Elisha Gray turned in at it, and got out the first machine. Bell's and mine came out about the same time. The machines are different. Bell's is what is called the magneto telephone, and mine the carbon. Those kerosene lamps that you see smoking yonder are my carbon manufactory. I peel it from the shades, and press it into buttons for use in my telephone. Were it not for my deafness, I would have discovered the telephone eight months before. While trying an experiment my deafness led me off on a wrong tack, and I was sloshing around on a false scent for months. But I have produced a good instrument. I have whispered into it here at Menlo Park, and been answered in a whisper by Henry Bentley in the Western Union office at Philadelphia."

Here the clock struck 3, and we started for the train. The Professor returned to his machine like a delighted boy, and as we left the house we could hear him gravely asking:

"How far is it from New York to Albany, from Albany to Syracuse, from Syracuse to Buffalo, from Buffalo to Cleveland, from Cleveland to Columbus, from Columbus to Cincinnati, from Cincinnati to Louisville, from Louisville to Nashville, from Nashville to—"

and so on ad infinitum till we were beyond hearing.

The Sun.

2. THE NAPOLEON OF SCIENCE.*

EARLY DAYS OF THE MARVELLOUS MAN AT MENLO PARK.

CALLED A LUNATIC AND DRIVEN OUT OF MEMPHIS–SNOWED UNDER IN CANADA–ARRIVAL IN BOSTON–HIS FIRST NIGHT'S WORK–THE BEGINNING OF A VERY WONDERFUL CAREER.

The marvellous discoveries of Prof. Thomas A. Edison of Menlo Park, N. J., have excited universal interest. His stock indicator, automatic and duplex instruments, telephone, electro-motograph, airograph, electric pen, and, above all, his speaking machine, mark him as the Napoleon of inventors. Indeed, at the Professor's age, Bonaparte had barely reached the rank of First Consul. As any particulars concerning the history of this extraordinary young inventor must prove of more than usual interest, the writer details a conversation with Mr. George S. Stewart, better known as Fattie Stewart, an old telegraph operator, now employed in the office of the Associated Press:

"I first knew Tom Edison," said Mr. Stewart, "in 1866. At that time I was an operator in Tennessee. Tom was employed by Col. Coleman, the Superintendent of the Western Union office in Memphis. He was a gawky boy, about eighteen or nineteen, and was reading everything about electricity that he could pick up.

* From the New York *Sun*, Sunday March 10, 1878, p. 6; clipping in Cummings scrapbook 4, NYPL.

He had a lean and hungry look, and always seemed to be under the influence of some secret excitement. He had got into his head the idea of sending duplex despatches, and all his spare time was devoted to experiments in the office. Coleman stood it for some time, but at last began to growl. He allowed that Tom was crazy, and said that 'any damned fool ought to know that a wire can't be worked both ways at the same time.' He declared that he wouldn't have Tom puttering around the office with such silliness, and finally discharged him in disgust. The boy went back home to some town in Michigan, and I lost track of him.

"Some time afterward I was transferred to the Boston office. At that time, wire No. 1, as it was then called, was considered the crack wire of the country. The fastest men were working it. For some cause the operator in Boston resigned. It was difficult to find a man to take his place. A half dozen fellows tried it, but found it too much for them. One after another, they dropped it like a hot potato, and sloped wiser than when they came. There was a man in the office named M. F. Adams. He thought the world of Tom Edison, and recommended him for the place, vouching for him as a first-class operator. G. F. Milliken, the manager, telegraphed to the little town in Michigan, asking Tom if he would come on and accept the position. Tom answered yes, and without further words started for Boston, via the Michigan Central and Grand Trunk Railroads. In running through Canada, he got snowed under, and was kept on the track in one spot for twenty-four hours, cold and hungry, without a bed. As usual, he owned but one suit of clothes, and that was on his back. Unfortunately, it was a summer suit. He might have frozen to death had he not bought an old rough roundabout overcoat, from a Canuck railroad laborer. But he finally got through all right.

"I was in the Boston office when he arrived, and I must say," continued Mr. Stewart, bringing his fist down upon the table, "he was the worst-looking specimen of humanity I ever saw. The modern telegraph tramp isn't a marker. He wore a pair of jean breeches six inches too short for him, a pair of very low shoes, the Canuck jacket, and a broad-brimmed butternut hat, a relic of his life in Memphis. The wide rim was badly torn, and hung down so that you could see his ear through the opening. There was the slightest trace of dirt on his upper lip, that he called a moustache. His hair hadn't been combed for a week, and he wore the blackest white shirt that was ever seen on the back of a human being. Nervously pinching his upper lip—a habit that he had—he inquired for the manager, and was sent to Milliken.

"Are you the boss?" Tom asked. Milliken smiled and said he was manager. Tom then introduced himself and asked when they wanted him to go to work. Milliken stared at him as though he couldn't believe his ears, and said 'At half past five.' It was then well along in the afternoon. Tom began to look around the office for a clock, and Milliken said: 'Young man, you have to work a pretty heavy wire.' Tom gave what he called his moustache an extra twist, and with all the assurance in the world blurted out, 'All right, boss. I'll be here at half past five.' He sloped so quick that it made Milliken's head swim.

"The operators burst into a peal of laughter. They had seen and heard everything and their remarks were anything but complimentary to Tom. 'Oh,' said one of them, 'he won't last as long as that Jerseyman that tackled the wire the other day.' 'Why, that fellow can't read by paper, let along by sound,' shouted another. A third declared that Tom was 'the worst he ever saw,' and when a fourth wondered 'whether the walking between Michigan and Boston was very good' there was a general roar.

"Well," continued Stewart, "half past five came, and so did Tom.

Everybody was on the *qui vive*. Milliken was just taking from the vault the supply of blanks for the night operators. As Tom came up he pointed to a pile of them, saying, 'Take what blanks you want and I'll show you your table.' Tom innocently picked up the whole bundle, and followed Milliken to his table. The operators began to grin and snicker. They all thought that he would get bounced after trying to catch one message. It was the No. 1 wire to New York. Jerry Borst, then considered one of the fastest senders in the country, worked the New York end. As Tom seated himself he heard the call 'B.' and turning to Milliken asked if that was the call for Boston. 'Yes,' replied the manager, watching Tom's movements with intense curiosity. Thereupon, Tom opened his key and ticked the answer, 'I, I!' Jerry began to 'whoop 'em up' in his best style and every eye was turned on Tom. He displayed no anxiety, but kept right along at his work as though he had been taking Jerry all his life. For four mortal hours did Jerry keep it up a hundred pounds to the square inch, and four mortal hours did Tom take it down in a handwriting as neat and plain as reprint. For the first time in his life Jerry had rushed it until he was tired without a break from the receiver. He was astounded. When he had finished, the following messages passed between them:

From Jerry,

> *Who the devil are you, anyhow?*

From Tom,

> *I'm the new man. My name is Tom Edison.*

From Jerry,

> *Well, by [a ripper—REP], you're the man I've been looking for for the last ten years, and you're the only man I ever found that could take me without a break. Shake.*

"And they shook. The astonishment of the boys in the office was unbounded. There was no more jibing nor snickering. Everybody was Tom's friend at once. The next day Milliken picked up a sheet of Tom's manuscript, and reflectively stroked his long beard. 'I never saw such pretty copy,' he said. 'He's as good an operator as I ever met.'

"At the close of the first night's work, Tom's friend Adams took him home with him. The first question asked was: 'What kind of a man is this Milliken? Do you think he'll let me experiment in the office when I'm not on duty?' Adams replied that Milliken himself was somewhat of an inventor, and he thought that he would not only let Tom experiment as much as he pleased, but that he would also take a personal interest in his experiments. The very first trial was the duplex despatches that gave Tom the reputation of a lunatic in Memphis, and caused him to lose his situation. Milliken, unlike Coleman, entered into the spirit of the thing, and in a short time Tom had so far perfected it that he worked it quite successfully between New York and Boston. But to accomplish this he spent every dollar he earned for material for his experiments, and when the grand secret was discovered hadn't money enough to pay for filing a caveat for a patent."

Stewart says that many persons witnessed Tom's experiments. Among others he mentions James G. Stearns, then President of the Franklin Telegraph Company. He appears to have dropped upon Tom's secret, and he had money enough to carry out Tom's ideas. At all events, he got a patent ahead of Tom, and reaped a large proportion of the benefits. Today his instrument is used extensively in this country and in Europe, and he is worth hundreds of thousands of dollars. Tom, however, got full credit for the invention of the duplex system through the news and editorial columns of the *Telegrapher*, a newspaper devoted to electric science, edited by J. M. Ashley, now of the *Journal of the Telegraph*. It was Tom's first newspaper notoriety, and he was greatly elated. He flourished a dozen copies of the paper over his head, and announced his intention of mailing them to Coleman, "to show him that the damned fool had actually succeeded in sending messages both ways at the same time on the same wire."

But Tom jumped from one invention to another, apparently utterly regardless of their pecuniary value. It was while he was in the Boston office that he invented the gold and stock telegraph indicator now in general use. In this case he pursued his experiment privately, and had money enough to get the invention patented. To-day it returns him a handsome royalty.

Fattie Stewart tells many amusing stories of Tom's career in the Boston office. His strange ideas and odd expressions gave the boys an inexhaustible fund of merriment. Pat Burns, now dead, was working nights in the Boston office, and attending Harvard law school in the day time. Burns was a magnificent operator, and was awarded Prof. Morse's gold key in the telegraphic contest years ago. He was a brilliant conversationalist, and passionately fond of argument. For the sake of it he was eternally getting up disputes with the boys in the office. Edison admired Burns's gift of gab, and when Burns was in the heat of an argument was wild to hear him talk. As Tom was quite deaf he couldn't catch the conversation at a distance. At such times he would disable his wire. His favorite method was to "ground" it under his table. While the chief operator was cursing and swearing and testing for the "ground," Tom would be off pulling at his upper lip and listening to Burns. The argument concluded, he would return to his table, take off the "ground" that the chief operator had failed to find, and innocently announce that the wire had come "O.K."

The Boston office was overrun with cockroaches, and Tom was much annoyed by them. With ready ingenuity he conceived and carried out a plan for their extermination. He tacked several zinc strips to the wall, at intervals of an eighth of an inch. He then applied the positive and negative poles of a battery alternately to the strips. He next smeared the wall above the strips with molasses. The roaches came up in platoons, very much after the manner of the British troops at Breed's Hill. As they stepped from strip to strip they "closed the circuit," received the full benefit of the electric shock, and dropped dead by scores. Tom used to catch their bodies in a water pail, and it is said that the bucket has been filled in a single night.

"Tom was naturally speculative in his ideas," said Stewart, "and the No. 1 wire kept him so closely employed during working hours that he hadn't any time for dreaming. One night he got into a discussion with the operator who worked the wire that connected with the old Atlantic cable at Plaister Cove. There was mighty little cable business in Boston, and Tom jeered the operator on his 'soft snap.' All he had to do was 'tend a repeater that was used only when the state of the atmosphere interfered with the working of the regular wire. The man was a first-class operator, and as he had got a little out of practice he thought a month's dash at Tom's wire would do him good. So, with Milliken's consent, they changed 'tricks.' Tom thus took the 'early trick,' from 1 to 8 A. M., and the cable operator took his place on No. 1. Tom found it more of a change that he had anticipated, for he was fond of lively company, and between the hours of 1 and 8 in the morning there were very few operators at work, and the office was as silent as the grave. So he went to dreaming in earnest. I can see him now sitting at his desk, pulling at his upper lip, and vacantly staring at the wall. His thoughts seemed concentrated on something beyond him— something apparently out of his reach. He's got there since, but he seemed to be a long distance from it then. After 2 A. M. he was left almost entirely alone. He was always somewhat musically inclined, and to relieve the monotony of the early morning hours he got some fine wire resembling the hair spring of a watch and attached it to his instrument in such a way that is sounded like an Æolian harp. And there he would sit through the long morning hour listening to this sad sweet music, utterly unconscious of what passed around him. Eventually, however, this novelty wore off, and he began to look for a fresh source of amusement.

"About this time," continued Stewart, "an order was issued that each night office at hourly intervals, between 1 and 8 A. M., should telegraph 'O. S.' to the New York Office to prove that each operator was awake and at his post. After his musical experiment became cold, Tom had fits of drowsiness, and, while indulging in a nap one night, the regular wire 'busted' east of Boston, just before the hour for answering 'O. S.' to New York. The chief operator at the New York end called 'Boston for test.' Poor Tom was fast asleep, and it was some time before he awoke. He found hell to pay. He very quickly substituted another wire for the one that had 'busted,' and was lucky enough to get out of the scrape with an admonition never to be caught napping again. But he took instant measure to protect himself and enjoy his naps. The office boy, Johnny McFarland, knew the 'call.' Tom took Johnny into his confidence, and Johnny promised to awake him on call. But the 'O. S.' business still cut him out of a square snooze, and he determined to get over the difficulty. He invented and attached a mechanical contrivance to the connections of the wire that would open and close the circuit and say 'O. S.' to New York, and sign his call 'B.' as regularly as the hour came round. Young Johnny faithfully awoke him when he heard any one call 'B." and after that Tom slept as sweetly as an infant.

"Tom's working the 'late trick' as the boys called it, gave him the day and part of the night to himself. He rented a room on Doane street in the rear of the Western Union office, and spent most of his time experimenting with everything that he could get that had any relation to electricity. His room was filled with old relays, sounders, wire of every size, length, and description, magnets, repeaters, insulators, batteries, blue vitriol acids, and books on electricity. His right-hand man was his old friend Milt Adams. In those days Tom was so taken up with his experiments that he spent upon them every cent he could raise, and went so far as to wear a shirt a month to save the price of washing. It was in this office in Doane street that he perfected his gold and stock indicator, and I reckon got the ideal groundwork for all his inventions."

At the end of the month the cable operator "weakened," and Tom returned to his old wire. His inventions, however, proved so valuable and renumerative that he resigned his position in the Boston office, and came to New York, where he quickly took the front rank among electricians. Stewart went South and lost track of him, but frequently heard of his surprising inventions.

"Three years afterward," he says, "I met him in front of the *Herald* building. To my surprise, he wore a plug hat, but it looked as though it had been stolen from some procession on St. Patrick's day. He was glad to see me, and asked all sorts of questions about what the Southern operators thought of his discoveries. I told him they were overjoyed at his success. He told me that he had got married, and in comparing his situation with the position he held in Boston, exultantly pulled three bank books from his pocket, and showed them to me, saying that he didn't feel 'quite so poor now as when in Boston, pounding brass with old Jerry.'"

The Sun.

3. THE INVENTOR OF THE AGE.*

AN AFTERNOON IN THE LABORATORY OF PROF. THOMAS A. EDISON.

The Marvellous Inventor at Work—How he was Bothered by a Wicked Tribune Employee—Three Talking Machines by the Ears—The Great Musical Telephone—Military Manœuvres by Cork-Headed Needles.

The recent appearance of Prof. Thomas Alva Edison in Washington, and his article on the speaking phonograph, printed in the *North American Review*, have increased public interest in the inventor and his inventions. Any information concerning him is sure to be read with avidity. In Washington he was the honored guest of the most distinguished men in the country. Robert Fulton, Sir Isaac Newton, or Galileo would not have attracted more attention.

About six weeks ago a journalist visited Mr. Edison for the second time. He was accompanied by the Hon. John Kelly of the Tammany General Committee. The Professor was found in his large laboratory on the second floor of his manufactory at Menlo Park, N. J. He was surrounded by jars of vitriol, strips of tin foil, phials of chemicals, smoking kerosene lamps, pieces of telephones, old clay pipes, odd looking electrical machines, copies of the great daily newspapers, mechanical etchings, papers of smoking

* From the New York *Sun*, Monday April 29, 1878, p. 3; clipping in Cummings scrapbook 4, NYPL.

tobacco, electro-motographs, and sundry bits of machinery of unknown power. An open volume of Poe's poems lay beneath the largest phonograph. Three or four visitors were wandering around the room. They were all personal acquaintances. One was playing a small organ in the west end, a second was coughing in one of the speaking machines, and a third was yawning over the Professor's carbon manufactory. A fourth proved to be a scientist. He bent over a piece of tin foil, freshly torn from the cylinder of a phonograph, and carefully examined the tracings through a magnifying glass. He was securing a supply of matrixes, with the view of solving the mysteries of the vibratory alphabet. He believed that the air waves produced by talking into the metallic diaphragm of the speaking phonograph left a particular mark or letter in the tin foil for each labial, consonant, vowel, or diphthong uttered. These marks or letters are very minute. They resemble the letters of the Morse alphabet. Through the aid of a microscope the scientist thinks that each letter or sound may be known by its length, shape, and depth. The secret once discovered sheets of words could be written by a steel needle, and the speaking phonograph made to talk of itself, and not merely mimic sounds murmured in its metallic ear. The savant pressed the bright alphabetical sheets to his breast, and left the laboratory full of confidence. The Professor did not seem to take much interest in the work of the scientist. He made no speculations on the result, but closed his eyes, laughed, and shook his head. Apparently he had little faith in the theory.

Mr. Edison's shoes were covered with Jersey mud. His shirt was collarless. He wore a seedy frock coat. There was a charcoal mark on the side of his nose, and another on the flange of his left ear. His hair seemed to have been spread with a hayfork, and his general appearance was decidedly interesting. The picture was not that of an alchemist; it was more like an honest plodding apprentice to a machinist. He was very busy. New York and Philadelphia were about to experiment with his carbon telephone, and he was making a connection with the two cities. He sped about the laboratory with a quick tread conferring with his assistants, making and receiving suggestions, and giving the necessary orders. For minutes he stood at the end of the room, with one pole of the telephone at his ear and the other at his mouth, shouting, "How do you get it now? Mary had a little lamb, its fleece was white as snow—how do you get that?" Finally the wires were so arranged that one of the assistants declared that he could overhear the electrical palaver between the two cities, and take a hand in if it was thought best.

The Professor approached his visitors with a smile of satisfaction. A roll of white paper that had been soaked in a chemical solution was drying on a chair near a hot stove. The day was damp and chilly. Mr. Edison removed the strips and offered the chair to a stranger. The Hon. John Kelly then handed him a bunch of cigars. They were thrown on the work table and everybody invited to help themselves. After lighting one of them, the Professor stepped to the window, and directed the journalist's attention to a little pond in the yard.

"They tell me," said he, "that you are a great fisherman. I like a little of it myself. Come down some summer's day, and I'll go out with you. That pond out there is filled with Montezuma trout."

"What are Montezuma trout?" inquired the innocent journalist.

"Some persons call them bull heads, and others catfish," he replied; "but Montezuma trout is what I call them."

There was a merry twinkle in his eye as he turned from the window and drew a score of letters from his pocket. "That first article in *THE SUN*," said he, "was a graphic description of the phonograph. I have received letters from everywhere, saying that the writers had read it. Many expressed the opinion that you were a fool, or a damned scoundrel, or both. Some complimented your powers of imagination, and others advised me to set myself right and refute the allegations and the allegator. The general impression is that the article is a pure invention, but if not, they all want a machine. I hadn't any idea that people were so skeptical."

He placed a few of the letters in the hands of the journalist. One of them was from a college professor in New Haven. He termed the writer of *THE SUN'S* description "a common penny-a-liner in the incipient stages of delirium tremens," and advised Mr. Edison to deny his assertions. "You have a good reputation as an inventor among scientific men," the letter continued, "but this SUN article is calculated to injure you. The idea of a talking machine is ridiculous, but the article is so ingeniously constructed, and some persons are so ignorant of the first principles of science, that they will be apt to believe it true, unless you deny it."

A second letter was directed to "Professor Thomas A. Edison" in a bold, round hand. Edison pointed to the word "Professor," laughed, and slapped the journalist on the shoulder. "I haven't forgotten that I owe that title to your maliciousness," said he. "I'll repay you with interest one of these days. I have received over 800 of these letters, expressing opinions derogatory to your character. What would you say if I should print them?"

After a lunch Mr. Edison said, "Come over to this table, and I will show you something surprising—sent me by a scientific gentleman." He filled a glass with water. Into the water he threw a half dozen needles, with tiny cork heads. He then picked up a magnet, shaped something like a spike, and held it a few inches above the scattered needles. They immediately formed a hexagonal group, like the following:

As the magnetic spike was raised, the intervals were widened with military precision, and as it was lowered the intervals were lessened. Two additional needles were dropped in the tumbler. Under the magnetic influence of the spike the six original needles opened square and the two newcomers took their positions. The square became octagonal, thus:

No matter how small or how large the number of needles thrown into the tumbler, the ranks were broken, and new squares formed with mathematical exactness. As the spike was raised or lowered the squares were expanded or compressed. But a more surprising result was obtained when the opposite end of the spike was turned toward the needles. Each one ran away as though anxious to hide itself, and they rolled over and over in a line along the rim of the tumbler in an effort to escape.

The Professor turned to the Hon. John Kelly, and sat down before his large speaking phonograph. Its cylinder was wound with tin foil. An old employee of the *New York Tribune* had been repeating in the mouthpiece a portion of the Hon. Horace Greeley's lecture to self-made men.

"Halloo, here's something on the foil now," said Mr. Edison. "Somebody has been talking to the instrument. Let's see what it is."

He adjusted a strap and ran the cylinder by steam power. The ear trumpet was placed over the diaphragm, and the machine began to imitate the peculiar piping tones of the lamented founder of the *Tribune*:

> My fri-ends: If I had had the cho-ice of a lecture to deliver here to ni-ight, I should have cho-sen some other subject than self-made men-it-is-an o-old lecture, and one which I have frequently delivered. I have another and a much better lecture which I occasionally deliver. The subject is wi-it. There-is-no-wit-in-it, but I am to-old it is a very good lecture.

> The Pasha of Egypt is the latest convert to subsoil ploughing. I say Pasha, my fri-ends—they-used-to-call-him-so-when-I-was-a-boy. They have a new name for him nowadays—I believe they call him the Keydeeve, or-some-such-name.

The Professor himself laughed heartily as the drawling tones and sharp nasal accents rolled from the funnel. But his merriment was increased when his assistant turned the handle of an instrument on another table that drawled out the following:

> Ye-es. Whenever you want to pay me fi-ive dollars you can never find me; but if you want to borrow fi-ive dollars you find me mighty qui-ick.

> Is there anybody in this office connected with the daily *Tribune*? No? We-ell, bub, if you see anybody in this office connected with the daily *Tribune*, tell them, for Go-od's sake, to put as much election news in the paper as the *Evening Post's* got. And be sure and tell Sam Sinclair I'm going to Texas. Come, Stuart, let's go.

Mr. Edison indulged in a long chuckle, during which the fire in his cigar went out. "I see you boys have been having your own fun here," he said. The cigar was relighted. He tore the matrix from the large phonograph, and wrapped a virgin sheet of foil around the cylinder. Then he picked up the blue volume of Poe's poems, and tenderly strained the beautiful ballad of Annibal Lee into the mouthpiece. It came out in tremulous accents, but a third phonograph began another imitation of Mr. Greeley, and the effect of the two machines was irresistibly ludicrous. The machines said:

EDISON'S MACHINE.

It was many and many a year ago.
In a kingdom by the sea,
That there lived a maiden whom you may know
By the name of Annibal Lee.
I was a child and she was a child.
In this kingdom by the sea;
But we loved with a love that was more than love,
I and my Annibal Lee.
And neither the angels in Heaven above,
Nor the demons down under the sea,
Can ever dissever my soul from the soul
Of the beautiful Annibal Lee.

MR. GREELEY'S MACHINE.

Who made up the State election table this morning!
Oh, he won't do to edit the *Tribune* almanac! No, no, no!
He'll never do for the almanac!
Oh, there are so many blockheads around this office that I shan't
have any head at all upon my shoulders by and by—at all—at all.
Now, boy, keep all the bummers out of my room tonight. I'm
going to lecture before a Father Mathew society,
and I want to write. Don't let any one in.
I want my proofs. Ask Sam Walter for my proofs.

Mr. Edison has a powerful, sympathetic voice, and the phonograph's rendition of his recitation might have drawn tears from a boarding school miss, were it not for Mr. Greeley's querulous interruptions.

But this performance was supplemented by one still more ludicrous. The Professor set the machines by the ears. They began to call each other names. The visitors screamed with delight.

"You're a liar!" shouted one instrument.

"You're another!" exclaimed the second.

"You're both liars!" bawled the third.

"I never stole a sheep!" said No. 1.

"You lie; you did!" blurted out No. 2.

"And so did you; I saw you do it," said No. 3.

Then followed: "You lie!" "You lie yourself!" "You both lie!" and similar expressions, until Mr. Edison's visitors fancied themselves listening to a barroom row.

There is one singular peculiarity about the speaking phonograph. To produce exactly the same tone of voice the cylinder must be turned at exactly the same rate of speed as when the words were spoken at the mouthpiece. If it is turned slower the sounds come out in a low bass; if turned faster they run up into a shrill treble. This difficulty has been overcome by the Professor's improved machines. The bed is a flat circular plate, with spiral grooves, about as large around as a tamborine. There is no cylinder. The sheet of foil is placed upon the flat plate, the point of the phonograph playing above the centre of the groove. The plate is turned from left to right by a well-tempered mainspring wound up the same as the spring of a patent lever clock, thus securing a uniform motion when receiving and repeating sentences. By these machines Mr. Edison said that he could register 48,000 words on a single sheet. Subsequent experiments, however, have convinced him that 40,000 is nearer the mark.

"Kiralfy, the actor, was here last week," said the Professor. "He was in ecstacies over the phonograph. He wants me to make him forty or fifty large cylinders, which he proposes to take to Paris. After securing the finest musical performers, each one will be invited to play into different diaphgrams, and the notes are to be recorded by the cylinders. The greatest prima donnas, tenors, contraltos, and bassos are to sing, and their words and melody transcribed by the phonograph. After securing copper matrixes, Kiralfy proposes to bring the cylinders to this city, and put them on the stage of the Academy of Music. They can be run by a Baxter engine a whole evening with fifteen cents worth of coal, and the opera given as produced in Paris without a musician or a cantatrice. What do you think of that? Only imagine the orders that would be given between the acts. 'Touch up Patti with a little oil,' 'Grease Capoul,' 'Oil up the first violin,' and 'Tighten the screw in the oboe'—what a treat that would be for a New York audience. I tell you, Kiralfy has it on the brain, and says he shall go to work immediately."

While Mr. Edison was entertaining his visitors with the phonograph, his assistants put in order a musical telephone. The instrument was placed upon a table in the centre of the room. A telegraphic wire connected it with a mouthpiece far away on the lower floor. The machine upon the table was not much larger than a common stock indicator. The chemically-soaked strips of paper dried at the hot stove were wound around a little wheel, and carried over the surface of a drum, pressed up to the mouth of the instrument. This wheel was turned outward by a small crank. An assistant on the lower floor murmured the "Marseillaise" into the mouthpiece below, and Mr. Edison began to turn the crank. The music filled the vast laboratory. It arose and died away like the notes of a powerful Æolan harp, apparently controlled in volume by the Professor at the crank. This instrument differs greatly from Bell's musical telephone. The latter sings through an ingenious arrangement of reeds. The music of the Edison machine is brought out by the action of a current of electricity upon a solution of sulphate of sodium in which the strips of white paper are soaked. The preparation of this paper and solution is the Professor's own secret. The paper goes into the instrument as white as snow, and comes out with a tinge of blue. Mr. Edison claims that a low murmur at the other end of the wire may be increased in volume by proper instruments until the effect would be deafening. He speaks of tearing roofs from music halls, and evidently has great faith in the scope of the invention.

Mr. Edison next exhibited his electro-motograph. It is claimed that this instrument would have given a perfect system of telegraphy were magnetism never discovered. In experiments Mr. Edison learned that certain chemical salts lost their friction when subjected to electric currents. A strip of paper moistened in a common chemical solution is laid upon a metal plate connected with one pole of a battery. A flat platinum strip fastened to the other pole is pressed down upon the moistened paper by the thumb. The thumb can draw the platinum over the paper as though it was greased until the electric current is interrupted. Then the normal friction of the paper is restored, and the hand will be involuntarily stopped. The elements of this discovery are applied to Mr. Edison's musical telephone, mentioned above. In that instrument the paper is carried over the surface of a drum touched by a steel spring with a platinum face. A sounding-board is stood upon edge, with a brass arm reaching to the centre of the drum. When the drum is turned the chemical paper is drawn under the platinum spring and the sounding board drawn outward by the friction. The electric current passes through the paper, destroying the friction, and the sounding board drops back to its first position. The cessation of the current again causes friction, and the board is again drawn out. This mechanical, chemical, and electric action is so instantaneous that the finest and highest notes of the female voice are reproduced 200 miles away.

Mr. Edison is 31 years old, and boyish in appearance. A distinguished stranger called on him not long ago. "There he is, sir," said an employee, pointing out the great inventor. The stranger took a good look at him and said: "No, no I want to see the old man." The Professor roared, and the stranger was so dumbfounded that he was hardly able to explain his business.

Mr. Edison's handwriting is wonderful. He forms his letters as they are printed and his communications bear such a close resemblance to the reprint that many persons deny that they are written. By this formation of letters Mr. Edison claims that he does not make any unnecessary marks, and that he is able to write much faster and more legibly than other persons.

The inventor is a great reader. He has a rare knowledge of English literature, and is thoroughly posted on the news of the day. He is a keen logician, enjoys a pithy joke or anecdote, and can talk Sam Cox or Ben Butler blind on the merits of the silver bill. He admires the true goodness of Deacon Richard Smith, loves to read the sweet funereal poetry of George W. Childs, A.M., believes in the histrionic ability of the Count Joannes, revels in the alleged veracity of Eli Perkins, takes a little stock in the peanut psychology of George Francis Train, and is down on all frauds, however high in station. His favorite expression in the presence of SUN reporters is, "Reduce the army to ten thousand men."

The Sun.

4. THE MAGICIAN OF SCIENCE.*

PROF. THOMAS ALVA EDISON GIVING A PUBLIC ENTERTAINMENT.

WHAT HE THINKS OF THE KEELY MOTOR AND THE GRINDSTONE BUSINESS—HIS VISIT TO MOUNT ST. VINCENT AND EDWIN FORREST'S CASTLE.

Few of the passengers awaiting the departure of the 9:10 train from the Grand Central Depot yesterday morning were aware of the presence of Thomas A. Edison, the marvellous inventor. Mr. Edison was on his way to the Academy Mount St. Vincent for the purpose of exhibiting his speaking phonograph, telephone, and electric pen. The inventor rarely appears in public. A sister of Mr. James R. Heenan, one of his old Kentucky friends, is a pupil in the academy. Mr. Edison consented to appear as a mark of friendship for Mr. Heenan. Over twelve years ago they were both employed in a telegraph office in Louisville. Mr. Edison was accompanied by Charles Batchelor, his right hand man for ten years; S. L. Griffin, his secretary; Martin Force and George Carman, assistants; and Mr. McLoughlin, the agent for the electric pen.

The great inventor took a seat in the car directly behind Mr. Robert Bonner, to whom he was introduced. The conversation turned upon Mr. Thomas D. Jones's discovery in the dressing of millstones. Mr. Jones claims to have made over $1,000,000 in ten

---

* From the New York *Sun*, Friday May 31, 1878, p. 3.

months by his invention. Mr. Edison thinks that the figures are grossly exaggerated. "Mr. Jones," said he, "says that there is a secret about the application of the diamond quartz to the millstones. Now as he has taken out a patent I don't see how the secret can be kept. Besides, if the invention is as valuable as he claims, it appears to me that he could make more money in this country than in England. We produce more flour here, and have more mills, and larger ones, than those in Great Britain. I can't see the necessity for running over to London."

Upon being questioned concerning his opinion of the Keely motor, the inventor said that at one time he had made arrangements to visit Philadelphia and see the motor. Mr. Keely was willing to receive him, but not willing to allow a thorough examination of his machine. "The one thing necessary for me to know," said Mr. Edison, "he refused to impart, and without this information I might as well look at a pile of broken machinery as at the motor. I had intended to take with me my assistant, Mr. Batchelor, with the idea of taking turns in watching the motor while in operation, night and day, and am certain that I should have been able to discover whether there was any real foundation for the extraordinary reports concerning the power produced by one little drop of water."

By way of comparison, Mr. Edison told a story of a steam engine with a new motive power claimed to have been discovered years ago by a gentleman in Newark, N. J. A well-known capitalist in this city was carried away by the plausibility of the inventor and lost $60,000. One of Mr. Edison's friends was about to embark in the enterprise but was warned in time. Mr. Edison examined the machinery and was satisfied that the engine was run by hidden shafting. He noticed that the jarring of the machinery was in perfect unison with the exhaust of an engine across the street. Subsequent developments bore out his theory. The inventor disappeared, and an examination showed that while the visible shafting was run by his engine, the engine itself was run by concealed shafting, the power being furnished by the steam engine across the street.

While Mr. Bonner was entertaining Mr. Edison with literary reminiscences the train stopped at Mount St. Vincent. The distinguished men shook hands and parted. On alighting, Mr. Edison was received by a priest, who conducted him and his party to the academy. The inventor wore a high silk hat. He said it was a new hat, but would not acknowledge that it was purchased for the occasion. He wore a plain turn-down collar, and a common black tie, a brown coat, and a pair of pepper-and-salt pantaloons. The good sisters in charge of the institution received him with great cordiality. Sister Marie and the good father showed him through the building, and introduced him to the teachers of the different classes. He was taken to the chapel, where the pews were filled with black-robed sisters, listening to mass. Though much impressed with the ceremony, the great inventor seemed sadly out of place. He was evidently unfamiliar with the service. While Sister Marie and the kind father knelt, he stepped behind a pew, put his hand to his ear, and endeavored to catch the words of the celebrant. He was lost in thought when the good father touched his arm as a sign to go, but the action was misconstrued by the inventor. He seemed to think it meant as a reflection on his lack of reverence, for he stooped as though about to fall on his knees, and remained with bowed head until his conductors left the chapel. Sister Marie showed him the twenty-six pianos used by the scholars in practising. They were in separate rooms in the second story. He seemed much surprised to learn that at certain hours they were all in use at the same time. From class room to class room he walked, with his hands behind his back. Rosy-cheeked pupils gazed at him in awe. Geological and astronomical charts were shown him, and the inspection was becoming monotonous, when he entered an apartment where there was a small steam engine. He turned the flywheel and caressed it as a man would caress a dog. All the scientific apparatus excited his interest.

After a thorough inspection of the academy, the party were conducted to the Norman castle, built by Edwin Forrest over thirty years ago. Mr. Forrest bought the ground in 1842. While the castle was being built he spent two winters with his charming wife in a stone cottage near by. When the castle was completed he filled it with choice paintings and exquisite furniture, but the most of it was never unpacked. A great domestic storm broke over the head of the actor. He slept two nights in the castle sitting in an easy chair. His life was blighted, but his heart ever turned toward his beautiful residence. He visited it many times after the property fell in to the hands of the Sisters of Charity and left with them substantial proofs of his regard. "A few days before his death," says Sister Marie, "I sent him one of our catalogues. On the morning that his death was announced I received a letter from him enclosing a check for $500."

Mrs. Forrest visited the castle but once after the decree of divorce. The Sisters showed her over the place without knowing who she was. The castle is octagonal in shape. The rotunda was Mr. Forrest's picture gallery, and the parlors, dining room, Mrs. Forrest's boudoir, the tragedian's library, and other apartments opened upon it. Two pictures frames are imbedded in the walls, but the paintings have been removed. They represented nude figures, and were considered out of place after the property was bought by the sisters. The library is now filled with a fine mineral cabinet presented by Dr. Edmund S. F. Arnold, a wealthy Protestant.

Mr. Edison and party were shown what is called the Cardinal's room. It is occupied by Cardinal McCloskey, who annually spends a few summer weeks in this delightful place. A painting of Peter's escape from prison adorns its walls. The apartment is richly furnished. A costly Bible lay upon the table with a volume of Father Lacordaire's sermons. After inspecting the castle Mr. Edison's party was invited to dinner in a private parlor. A hum as of a hive of bees arose from below. Over two hundred girls were dining and talking in the refectory beneath. On Thursdays and Sundays they are allowed to talk during meals. On all other days the strictest silence is maintained. Mr. Edison was seated at the table opposite a painting 300 years old, much admired by Cardinal McCloskey. It is a representation of the Child and Virgin. During the visits of the Cardinal a seat at the table is reserved for him, where he can feast his eyes with the beauties of the picture. One of the most striking paintings in the institution is an illustration of the work of the Sisters of Charity in Paris. It is valued at several thousand dollars, and was presented to the academy by the Hon. John Kelly.

During dinner Mr. Edison spoke of his recent trip to Washington. He visited the White House with a phonograph at 1 in the morning, and was cordially greeted by Messrs. Hayes and Schurz. Mrs. Hayes had retired, but on hearing that the wonderful inventor was downstairs she arose and quickly joined the party. Every attention was paid Mr. Edison, but as he wanted to catch the 2 A. M. train he was compelled to decline extended civilities. He spoke of Mrs. Hayes in the highest terms, and turning to the SUN reporter smilingly said: "I scanned her husband's forehead very carefully, but the letters were not there."

After dinner a phonograph was placed upon a table on the platform of the lecture room. The sisters filled the galleries, and invited guests and misses attending the institution were seated in the hall. Mr. Edison ascended the platform and was greeted with a round of applause. He was somewhat embarrassed, but made a half bow to the audience, and disappeared in a room at his left. He was told that he was expected to make a speech, explaining his remarkable discovery, and giving an idea of the uses to which it could be put. The proposition seemed to take away his breath. He declared that it was impossible. He had never made a speech in all his life, and would not dare to attempt one. Thereupon Mr. Gus J. Thebaud took the platform, made a humorous speech, and introduced the Professor. Mr. Edison sat down with his back to the audience, put his lips to the mouthpiece, and in a clear, sonorous tone recited the following:

The dying soldier faltered, and he took that comrade's hand,
And he said "I never more shall see my own, my native land.
Take a message and a token to some distant friends of mine,
For I was born at Bingen, at Bingen on the Rhine.

The Professor then readjusted the instrument, placed a large paper funnel over the mouthpiece, pointed it over the head of his audience, and turned the crank. The words were repeated perfect in tone and accent, and were distinctly heard in every part of the lecture room. Mr. Edison then whistled "Yankee Doodle" and counted twenty-five over the same matrix. They were all reproduced together as follows:

BINGEN.	YANKEE DOODLE.	THE COUNT.
The dying soldier faltered, and he took that comrade's hand,And he said "I never more shall see my own, my native land. Take a message and a token," & c.	Whoo, whoo, whoo whoo, who- who-who, whoo-whoo-whoo- whoo-whoo, who, who, Whoo-whoo-whoo, whoo, who-who-who, who, who, who, who, who, who-who!	One! Two! Three! Four! Five! Six! Seven! Eight! Nine! Ten! Eleven! Twelve! Thirteen! Fourteen! Fifteen! Sixteen! Seventeen! Eighteen! Nineteen! Twenty! Twenty-one, &c.

A storm of applause followed. The Professor then talked to the machine thus:

There was a little girl,
And she had a little curl
Right in the middle of her forehead.
And when she was good
She was very, very good,
And when she was bad she was horrid.

It came out of the machine so perfectly that the audience roared with laughter. The phonograph then sang in a clear tone:

John Brown had a little Indian,

And followed it up by whistling:

Oh, happy, happy, happy be thy dreams.

After talking French, laughing, coughing, imitating the cackle of a hen, and winding up in a broad Yankee accent with the words, "Wall; I do declare!" the inventor made a jerky bow, and retired amid applause. Mr. Batchelor took his place, and kept the spectators in extraordinary good humor.

The telephone was then exhibited. Through the use of the paper funnel persons singing in a room 150 feet away were heard by the audience. A music box played at the other end of the wire was also audible. The instrument was a common telephone, and should not be confounded with the musical machine, which is much more wonderful.

Mr. McLaughlin then exhibited Edison's celebrated electric pen. A wire is attached to the pen or stencil, and while a person is writing, a steady stream of electricity perforates the paper, making almost invisible little holes corresponding to the formation of the letters. As you write the needle is projected into the paper at the rate of about 8,000 punctures a minute, forming perfect autographic paper stencil. The stencil is then secured in a frame or press, a felt roller saturated with printer's ink is passed over the face of the stencil, and the perforations become filled with ink, which is deposited upon the paper underneath. From 1,000 to 15,000 copies can be taken from one stencil prepared with this pen.

This closed the entertainment. The good sisters warmly thanked the great inventor and he ran for the cars. It rained, and his new silk hat was getting wet. He reached the city in safety, and left for Menlo Park on the 5 o'clock train.

The Sun.

5. EDISON'S "EAR TELESCOPE."*

THE RELIEF THAT THE INVENTOR IS PERFECTING FOR THE DEAF.

CONVERSATION CARRIED ON WITH EASE BETWEEN PERSONS TWO MILES APART—A NEW DEVICE THAT WILL ENABLE DOCTORS TO DETECT DISEASE.

Upon his latest discovery, which he hopes to utilize principally for the deaf, Edison has not given much thought for a few days. He has had something else in mind. An unusually large number of persons called on him at his laboratory in Menlo Park yesterday, and they found him very busy upon experiments with carbon. His hands were as black as soot and there were smudges of grime all over his face. With an alcohol flame, into which he thrust little tin plates, he converted different substances into carbon, and then tested the resistance and heat, detecting the power of each kind of carbon with the galvonometer.

On one side of him a workman talked nursery rhymes into a phonograph as a matter of business, for the tin-foil record is to be used for a toy phonograph, large quantities of which are wanted for the trade as soon as they can be perfected. These will be sold for two or three dollars each, and will repeat "Mary Had a Little Lamb," "Jack and Jill," "Yankee Doodle" whistled, and a nursery hymn sang.

---

* From the New York *Sun*, Saturday June 8, 1878, p. 3.

Edison looked very tired yesterday, and said that he was, but he was feeling very happy at the success of an experiment with the carbon telephone the night before. He had fixed up his fine telephone at the end of the laboratory, and Mr. Bachelor, his assistant, held the other end of the wire in his house, a quarter of a mile away, but there was a resistance of a hundred miles in the wire. That is, the effect was the same had Mr. Bachelor been in Baltimore and Edison in Menlo Park.

Without any funnel to concentrate the tone, Edison said:

"Now Bachelor, I am going to walk to the end of the room and talk to you."

Mr. Bachelor heard him as plainly as did Mr. Griffin, who stood by Edison. He also heard Edison walk across the room. Then Edison in his ordinary tone said something. The telephone caught it up 150 feet away and sent it through 100 miles with perfect distinctness to Bachelor.

"That was the most perfect piece of work ever done with the telephone," said Edison yesterday.

Leaning against the table that supports the telephone were two immense funnel-shaped things that looked like confectioners' cornucopias, big enough to hold a ten-year-old boy.

"These," said Edison, "are what I made my experiments with for the megaphone. I see some of the papers speak of it as a telescope-phone. That isn't the name. I guess Redpath, who saw me experimenting with it, called it that by way of a joke. That's a megaphone—great sounder—but it isn't done yet. I've got the principle, though. See here." And Edison lifted one of the great paste-board funnels so that the large end rested on the table. "Now, put your ear to that," he added, pointing to the little end. "You hear things outside plainer than you did before."

Indeed it was so, and so plainly as to torment the ear.

"But that one is no good," he said, "for this purpose. This is the one, and I have got the principle in it that I was looking for."

He tossed aside the biggest funnel, and took up one somewhat smaller. It was made of thick cardboard, bound with iron hoops like a barrel.

"That's about all there is to it," he said.

"But what is it going to do?" the reporter asked.

"Well, I wanted some simple, convenient thing that deaf people could use in a theatre, church or anywhere. I had an idea that I could, if I could get just the right means for catching the vibrations of air that produce sound. I could carry them into a little box or chest that might rest in a man's lap as he sat. From that chest tubes might run concealed in his clothing to his ears. The point was to find out just how to catch the sound vibrations."

Then Edison got on the table, and, swinging his feet, began to tell how he first reasoned out his theory, and then experimented to prove its truth; and Mr. Griffin, his old-time fellow-operator, friend, and recently engaged assistant, says that the narration simply illustrated the mental operations and subsequent investigations that have led to all his inventions.

"You see, sound is produced like light," Edison went on, "by vibrations. Well, what do you do when you want to get light from a distant object, or if your eyesight is bad, to see clearly something near? Why, you make such aids to the eye as will catch and carry to the eye the vibrations of light. When you have done this you see clearly. Well, now it struck me that you'd got to treat the ear in the same way, when anybody's deaf or wants to hear something away off. But the telescope isn't good for much unless the object glass which catches and focuses the light is just at the proper angle. What we want for the ear, then, on the same principle, is an instrument the line of whose angle is exactly right. Then you are going to get your sound vibrations, because they will be caught and focused. The fact is, I'd got to get an ear telescope. Then we made that one," pointing to the big funnel, "but it wasn't right. 'Twouldn't work. Then we made the other one."

"You didn't make it at haphazard, did you?"

"Why, no. I had the general idea of the angle required."

"You mean by the angle, the angle of divergence of the sides of the telescopic tubes from its centre?"

"Yes, that's it. With the proper angle sound vibrations are received, and then conveyed without break or interruption to the ear."

"How did you find that this was right?"

"You see that barn over there?" pointing to a barn a mile or so distant. "Well, I held this tube to my ear, and I heard a man talking over there perfectly. Just as the telescope catches the rays of light from a star and brings them to a focus, so by pointing this tube directly at the spot that the sound comes from you catch the vibrations of sound."

"Then this does not magnify sound?"

"No; only concentrates it. Now, this angle is all there is to this. I take and cut off a section a finger's length from the end and so I get a short tube with the proper angle, suitable for a comparatively short distance. If I have enough of them they will answer as well as one big tube. Now, perhaps I can carry out my idea and make a little chest, from which these tubes will issue. They will catch the vibrations that are made when a man speaks or preaches, and carry them into the chest, and then they'll pass through the tubes to the ears. I don't want to say that they will do it, because I haven't made the thing yet and tried."

"You think that with such a thing deaf persons will have no more trouble in church or at a theatre?"

"Well, that is what I was thinking about."

For long distance conversation the instrument will have a different shape as a matter of convenience, and be provided with a speaking tube. There will be two of the ear telescopes, and to the smaller end of each a rubber tube will be joined, for convenient contact with the ear. Between the two telescopes, and connected with them by braces is the speaking tube, which has a very slight angle of divergence. Thus, two civil engineers a mile apart wish to talk. Each of them has a megaphone. Placing the tubes in the ears, one of the engineers can talk through the speaking tube in an ordinary tone of voice, and the other will hear and reply. The instrument is different from the aerophone, which Edison has thought of, the utility of which is for very loud talk at very great distances. Edison proposes to rifle the speaking tube of the megaphone so that the vibrations may be kept as compact as possible.

Already some extraordinary experiments have been made with the rough cardboard ear telescope, and its utility for the purposes of ordinary conversation for a distance of two miles was shown at Menlo Park a few days ago. It requires no shouting. Even a whisper can be heard 200 yards away.

"But do you mean to say that you require no electricity, no carbon, and no connecting link?" asked a Newark doctor who had stood by as Edison explained the invention to the reporter.

"Oh, no; there is nothing to it but the making of the tube right."

"How does it differ from the aerophone?"

"Why, that opens valves, you know, by compression of air, for the purpose of throwing sound over a great radius. This concentrates vibrations instead."

For field and long distance purposes the megaphone will be mounted on a three-legged table like a surveyor's instrument. For audience rooms it may rest upon the lap, and, of course, has no speaking tube.

Already letters pour in upon Edison from deaf persons who have heard that he is at work at something that will give them ears as spectacles give eyes to others. Sometimes he receives as many as 100 a day, and he has had a batch of circulars printed, for it would take his secretary all day to answer the inquiries. These are mailed to every one who writes.

Edison laughs heartily when he refers to Joseph Medill's letter. The editor of the Chicago *Tribune* said that he would give $10,000 for something that would help his deafness, and he believes that he has got something that will do it. At the same time he is so cautious that he will not predict success for the lap megaphone simply because he has not made and tested one. The field megaphone he knows is a success.

To-day Edison begins experiments on another invention. He will try to combine the principles of the telephone and microphone so that physicians may have an instrument that they can use instead of the stethoscope.

The Sun.

6. EDISON'S THUNDER STOLEN.*

SOME ENGLISH ELECTRICIANS LAYING CLAIM TO HIS DISCOVERIES.

SECRETS OF THE MENLO PARK LABORATORY SEIZED UPON IN EUROPE—THE YOUNG INVENTOR EXPOSING THE PRETENDERS— HOW IT WAS DONE.

The European mails of this week bring full reports of the address delivered on the 23d of May at a special general meeting of the Society of Telegraph Engineers, in London, by Prof. W. H. Preece, on "The Connection Between Sound and Electricity," illustrated by the alleged recent discoveries of Prof. Hughes. As reported in the London *Times*, Prof. Preece said that during the last few months the science of acoustics had made marvellous and rapid strides. First, they had the telephone, which enabled them to transmit human speech to distances far beyond the reach of ear and eye; then they had the phonograph, which enabled them to reproduce sounds uttered any place and any time, and now they had, he might almost say, a still more wonderful instrument, which not only enabled them to hear sounds which were otherwise absolutely inaudible, but also enabled them to magnify sounds which were audible— he alluded to the microphone, an instrument which acted toward the ear in the same capacity that the microscope acted toward the eye. In this new instrument, too, diaphragms were cast aside. The

* From the New York *Sun*, Sunday June 9, 1878, p. 7.

instrument was the invention of Prof. Hughes, and it was made with common nails, a common piece of wood, a halfpenny money box, and common sealing wax, and by this apparatus he had been enabled to attack nature in her strongholds.

On the same topic, the London *Lancet* says:

Dr. Richardson has been experimenting during the last week with Prof. Hughes, the distinguished discoverer of the microphone, or sound magnifier, in order to ascertain whether the apparatus can be applied to diagnostic purposes in auscultation of the lungs and heart. It has been assumed that an instrument which makes the steps of a fly audible at a long distance can at once render any movement in the human body acutely perceptible through the sense of hearing. This is, however, a premature assumption. By means of the microphone Dr. Richardson has been able to detect the respiratory murmur and the sounds of the heart, but not practically better, up to this time, than with the stethoscope simply. The fact is that there is considerable difficulty in making the sounds which are heard through the stethoscope pass in the form of electrical vibrations to the ear, and until this is achieved the application of the microphone for the purpose of diagnosis is not available.

It may be remembered that upon the telegraphic announcement of these discoveries, Mr. Thomas A. Edison, the inventor of the phonograph and the telephone and over marvels, pronounced the pretentions [sic] of Prof. Hughes a deliberate "steal" of his inventions.

A SUN reporter showed Mr. Edison the accounts in the London papers quoted above, and asked, "What do you say to that?"

The inventor stood in the doorway of his laboratory, dressed in his working costume, and wearing his historic slouched hat. His hands were still grimy from contact with tools and machinery of the workshop. He glanced rapidly over the articles. His first exclamation was:

"Phew!"

Then he read a little further and ejaculated:

"Well!"

At length, as he finished the perusal, he said with his most comical of expressions, and with a twinkle of humor in his eyes: "I declare that this is the coolest, cleanest steal that I ever knew. This man talks of this thing as thought it was entirely new, and as though he believed it was the invention of Hughes, when he has the most positive evidence that the thing is mine. But come down to the house and have some supper, and I will talk to you about it afterward."

In the house, alternately devoting his attention to supper and to Tom Edison, Jr.'s efforts to consume the entire bowl of sugar, Mr. Edison continued: "The man Preece has had my telephones over three months. He is Electrician of the British Postal Telegraph. He visited my laboratory in June or July, 1877, and I showed him freely everything I had done up to that time. There was no reserve, but he examined all my plans, drawings, and machines."

"Did you show him the microphone?"

"Why, of course I did, because the microphone is contained in the telephone; it is nothing but a finely-adjusted telephone. To say that the microphone is a superior invention to the telephone is absurd, because it is only a part of the telephone. There would be no use in adjusting a telephone to such a delicate pitch, because the jar of a building, the hum and roar of the city, would keep up a continual buzz. Hence it could not be a practical articulating telephone. You have got to reduce the sensibility of it to make it a practical thing. You can see for yourself that if a machine should be registered to record the sound of the step of a fly, the sound of the human voice would be a perfect roar in it, because the sound of the voice is thousands of times louder than the sound of the step of a fly.

But the telephone can be adjusted so as to record the sound of the step of a fly, as they say the microphone does. The base principle of the whole thing, however, is due to my undisputed discovery of the fact that certain substances called 'semi-conductors,' such as carbon, various oxides and sulphides, vary their resistance to the passage of the electric current by pressure. In July, 1877, I filed a patent, which is now issued, for an apparatus, which is, when properly adjusted for transmitting the sound of the voice, a telephone, but when adjusted delicately, which can be done by simply turning a screw, it is a microphone."

"Was this carbon principle shown to Prof. Preece in your laboratory?"

"Certainly; it was shown to him in the telephone, in which the direct impact of the sonorous vibrations was used. This was called a pressure relay,' an account of which was published in the *Journal of the Telegraph* in July, 1877, and a month or so afterward in the London *Telegraphic Journal*. It was to be used as a repeater or transistor from one circuit or telegraph line into another telegraph line, of the telephone vibrations, thus allowing conversation to be carried on between points widely separated. Mr. Preece was also shown by me over two hundred different combinations of one material and another with carbon; also the effect of pressure on the passage of the current, and one of the devices described by Hughes was precisely the same in form and principle, as will be seen in the *Scientific American* and the supplement of June 8, 1978.

"While Mr. Preece was here he suggested that after I had my telephone completely perfected, if I would place it in his hands he would bring it out in England. From time to time I kept him posted about all my experiments. My principal difficulty here, previous to my sending him my telephone, was due to the expansion of the telephone case. The mere heat of the hand expanded it. Great difficulty was experienced by us in removing this defect. But we did it by a very simple contrivance. This also is claimed by Hughes in his latest publications. He pretends to have discovered the use of this carbon as a measure of heat—the expansion of the tube that holds it as a heat measure. Why, I sent Mr. Preece a copy of the Washington *Star* containing an account of the carbon telephone used as a very delicate heat measure."

"Is there any evidence that Mr. Preece was in communication with Prof. Hughes and told him of your inventions?"

"Evidence! Why, I should say there is the very best kind of evidence. It is nothing less than Hughes's own acknowledgement in his first lecture before the Royal Society. Here is a slip of it. He says, 'My warmest thanks are due to Mr. W. H. Preece, electrician to the Post Office, for his appreciation of the importance of the facts I have stated, and for his kind counsel and aid in the preparation of this paper.' I should think he ought to give Preece his warmest thanks, for Preece has deliberately stolen my inventions to carry to him. At the same time Preece had in his possession a direct impact telephone of mine precisely similar in principle but far superior in practicability to this described and claimed by Hughes."

"Can you prove that?"

"Certainly. I sent over one of my assistants, Mr. Adams, with it, to deliver the telephones and do anything that Preece required. He was taken sick upon his arrival, or he might have shown up Preece's pretentions [sic]."

"But he did deliver the telephones?"

"Yes, and I afterward sent more telephones to Col. Gourand, in which the effect of expansion was entirely eliminated. In *Scribner's Monthly* for April, and the *Journal of the Telegraph* for April, and the *Journal of the Franklin Institute* for April are illustrated articles on the direct impact telephone, precisely the same in principle as that claimed by Prof. Hughes.

"I've got the documents you see. The idea of this man, having these things in his possession, having the impudence to make these claims for Hughes, beats me." And then Mr. Edison gave one of his jolly laughs, as if he really enjoyed the joke.

"Look here," he continued, taking a letter from his pocket. "Here is a letter from Preece to me, recently received, in which he says: 'Hughes's doings border very closely upon yours, and it is quite difficult to distinguish between what you have done.'

"I should think it would be," laughed Mr. Edison. "Under the circumstances I should think it would be quite impossible to tell the difference.

"Here is another extract from Preece's letter to me: 'I'll be very glad to bring out anything you have if you send it over.'

"I should think he was bringing out my things with a vengeance. As for Hughes, I know him very well. He has made money by patents. He says in his Royal Society lecture: 'I do not intend to take out a patent, as the facts I have mentioned belong more to the domain of discovery than invention.' What has he discovered? I should like to know. The whole principle of what he claims to have discovered was published by me over a year ago. Why, I obtained a patent in England six months ago for what he pretends to have discovered! I have sent papers containing the accounts of my inventions, covering what he claims, to the Count de Moncel, Paris, Prof. Schellen and Dr. Zetsche in Germany, and other scientific men in Europe, illustrating the progress of the telephone. Hughes can buy a copy of my English patent for ten cents that will show how foolish are his pretences to the discovery of the microphone. If you take the shilling piece and this nail business that he talks about, it will give sound, but it will not represent the original sound. By means of the carbon we reproduce the original sound, the scraping of the finger nail, the rustling of the paper, &c. The discovery that I have made and patented consists in finding some material that would transmit waves of electric current which should be proportionate to the sound waves. That was my discovery. It has been published in thousands of journals for the last two years, and in some of the very English scientific journals that now parade Hughes's discovery. They ought to read their own newspapers to see how foolish are Hughes's pretensions. Here, for instance, is my claim, published in this country, April 16, 1878:

> In the latest form of transmitter which Mr. Edison has introduced the vibrating diaphragm is done away with altogether, it having been found that much better results are obtained when a rigid plate of metal is substituted in its place. With the old vibrating diaphragm the articulation produced in the receiver is more or less muffled, owing to slight changes which the vibrating disk occasions in the pressure, and which probably results from tardy dampening of the vibrations after having been once started. In the new arrangement, however, the articulation is so clear and exceedingly well rendered that a whisper even may readily be transmitted and understood. The inflexible plate, of course, merely serves, in consequence of its comparatively large area, to concentrate a considerable portion of the sonorous waves upon the small carbon disk or button. A much greater degree of pressure for any given effort on the part of the speaker is thus brought to bear on the disk than could be obtained if only its small surface alone were used.

"Now, compare this with Hughes's claim, published in London on May 9, 1878, over a month later:

> It will be seen, however, that in the experiments made by myself, the diaphragm has been altogether discarded, resting as it does upon the changes produced by molecular action, and that the vibrations in the strengths of the currents flowing are

produced simply and solely by the direct effect of the sonorous vibrations.

"Compare again my invention as described in the *Scientific American*, as follows:

By constant experimenting, however, Mr. Edison at length made the discovery that when properly prepared, carbon possessed the remarkable property of changing its resistance with pressure, with Hughes's claim, published a month later, as follows:

> It is quite evident that these effects are due to a difference of pressure at the different points of contact, and that they are dependent for the perfection of action upon the number of these points of contact.

"I had arranged with Prof. Langley, Director of the Allegheny Observatory, in Pennsylvania, to use my carbon telephone as a heat measure, and an apparatus for that purpose was exhibited some six months ago, and at his request I was engaged to perfect a machine for measuring an infinitesimal amount of heat. This is to be used in determining the different degrees of heat in the solar spectrum, and the apparatus has been shown to many scientific men in my laboratory in the last six months. I experimented with it before a number of scientific men attending the Academy of Science at Washington, on the 19th of April."

The hour for the reporter to leave came all too soon, and as he went from the laboratory he heard Mr. Edison saying to his associate, Mr. Batchelor, "Let 'em steal the microphone if they will; that is only one little thing. Before another two years go by I'll give 'em phones and graphs enough to make 'em sick."

7. FOUR HOURS WITH EDISON.*

THE GREAT INVENTOR AGAIN REVELLING IN HIS MENLO LABORATORY.

A Tiny Bride's Visit—Old Telegraphic Chums Go to See Him—A Schutzenfest—Tales of the Border—More Wonderful Discoveries—An Ink for Blind Persons—Edison at Home.

At 4 o'clock on Monday afternoon Geo. F. Stewart, Edwin M. Fox, two old telegraphic chums of Prof. Thomas A. Edison, and Col. Michael C. Murphy sat upon the stoop of the laboratory at Menlo Park. The Professor was out riding with his wife and children. His factotum, Mr. Griffin, was escorting a tiny bride and her husband through the laboratory. There were but few changes in the long room. The chemicals had not been removed from their shelves, and the jars of green vitriol cast a lurid light upon the floor. The kerosene lamps occupied their usual places under an immense inverted funnel, and were sputtering and smoking while a boy was scraping carbon from their chimneys. The organ was out of tune, and the half completed machinery scattered about the room had a dusty look. A big board marked like a target, with a hold in the centre, stood upon one of the tables. It is to be used in the aerophone, which the Professor proposes to erect in front of his factory, and with

* From the New York *Sun*, Thursday August 29, 1878; clipping in Cummings scrapbook 4 NYPL.

which he says he can hold sweet intellectual conversations with old Bill Allen of Ohio. A half dozen improved speaking phonographs sat upon another table, near an open box of tinfoil. Shellac is no longer used in holding the foil on the cylinder. It is wrapped about the cylinder, its ends meeting over a groove, where they are fastened by a strip of steel which is pressed into the groove.

The tiny bride looked like a fairy in the den of an alchemist. Her black eyes dilated with wonder and her cheeks were flushed with excitement. Griffin got the phonographs by the ears, and they thumped the "Dying Solder of Bingen," whaled the "Little Girl with a Little Curl," and prodded "Mary's Little Lamb" without mercy. Meantime a stranger began to meddle with an electrical machine, and received a shock that gave him a back-actioned colic. He doubled up like a jack-knife, and, on observing a smile rippling over the face of the tiny bride, saw something through the window that attracted his attention for several minutes. The telegraphic operators, warned by his fate, stopped fooling with a magnet so powerful it could have picked up a small bar of railroad iron, and the tiny bride, robbed of a further source of amusement, left the laboratory.

The operators returned to the veranda, lighted cigars, tipped their chairs back against the siding, and were retailing spicy telegraphic reminiscences when a rockaway wagon appeared in the distance.

"That's Tom, now," said Griffin.

Mrs. Edison was driving. The Professor sat at her side with his hands pressed between his knees. His little ones were piled in behind. The rockaway stopped before the gate, and he sprang out. He wore a straw hat and linen duster. His moustache had disappeared, and he looked more like a boy than ever. Half way up the path he recognized one of the operators and shouted, "Halloo Stewart." Shaking hands all around, he backed up against a pillar, and chaff began to fly. It was the same spirit that animates bootblacks or brokers when they meet in a group and begin to guy each other.

Mr. Edison looked well. His face was bronzed, and his scalp lock stood up like a tassel on a stalk of broom corn. He detailed the incidents of his trip on the plains, told how he had coaxed caloric out of Arcturus and Vega with his tasimeter, and laughed over the astonishment of the astronomers when they took observations, figured away all night, and found that the town of Rawlings was three miles from the spot laid down in the Government maps. "The town must have been moved at night when the inhabitants were fast asleep," he said, "for the Government officials could not possibly make a mistake." He was satisfied with the tasimeter, and thought it might be used to measure the heat of stars so far away that they could not be seen through telescopes.

He described his visit to Yosemite. Arriving in the valley at 3 P.M., he rode to the top of Glacier Point, came down the trail after dark, and started for Mariposa the next morning. Horace Greeley was the only man who ever beat his time. Mr. Greeley reached the valley after dark, and left it before daylight. The Professor anathematized the Coulterville route, and was pleased with the spanking way in which he was ripped adown the mountain roads of the Sierras. A lady, weighing at least 250 pounds, was a fellow passenger. As the driver cracked his whip, and turned corners shelving awful chasms at full gallop, she braced herself, and when the danger was passed fervently ejaculated. "Thank the Lord." She was the Widow Van Cott, the revivalist. Her daughter accompanied her, and was delighted with the ride.

The Professor spent a day in Virginia City and one in San Francisco. He received much attention, and was overjoyed to meet in these cities old telegraphic operators with whom he had been intimate in his boyhood days. From Rawlings he went on a hunting expedition, and was charmed. "It was great," said he. "I went to sleep on the ground under the blankets one night, and when I awoke in the morning found my eyes full of water. It was raining, and the water had formed in pools over my eyelids while I was asleep."

When asked if he had seen any rattlesnakes, he replied, "No, I heard a great deal about them, but saw none."

Griffin suggested that he might not have had the right kind of whiskey, and the Professor laughed. He went into ecstacies [sic] over the hunting. Antelopes and jack rabbits covered the ground, "but," said he, "you must be a good marksman to hit them. Fox here was with me. He and I became pretty good shots, but Fox can beat me shooting round the corner." This allusion brought out an incident of the trip. Fox had been inveigled into shooting at a plate through a curved rifle barrel.

"You're right," he answered, "but I never had to go up and kick over a jack rabbit after I had shot it before it would fall." This developed another incident. The Professor had crawled a hundred yards on his hands and knees, flanked a stuffed rabbit skin, and filled it with bullet holes.

The guides and trappers excited the Professor's admiration. "One of them," said he, "trailed us eighty miles from Rawlings over a perfectly wild country, and delivered despatches [sic] received after we had left. I was lying down when they told me a man had come in with despatches. I thought it the foundation of some joke, and concluded not to bite. When he came up and handed me the letters I could hardly believe it. It was barely possible that he might have trailed us over the open level country, but we had crossed miles of flat rock, where we left not the slightest track. I asked him how he trailed us over the rock, and he said by the tobacco quids. And I guess he was right," continued the smiling Professor, "for we were all thoroughbreds. I don't believe there was a man in the party who didn't chew tobacco."

The trapper charged them only $25 for trailing them up and delivering the messages. He stayed with them until they returned.

On being asked what he had been doing since his return home, the Professor said: "Well, I'm going through my letters. I've set the men all to work, but the fact is, I'm a little off my centre, and haven't done much. I tried a thing that I had been thinking over for more than six weeks, but it was no good. I left out something at the beginning that I thought was of no account, and it turned out that it was just the thing that I ought to have kept in. The whole thing was a dead failure."

The Professor then described his visit to the Consolidated Virginia mine in Nevada. He made one important discovery. The miners hunt for the ore deposits with diamond drills. Great clay walls divide these deposits, and when one is discovered, thousands of dollars are spent in ascertaining its extent. Mr. Edison thinks that all this could be done in a few days by what is known to electricians as a "ground wire." He could ascertain the extent of the deposit by the amount of resistance offered to the wire. "A new deposit has just been discovered in one of the mines on the Comstock lode," said the Professor, "and they will probably spend $30,000 or $40,000 in finding out how large it is. I am sure that with a ground wire I could measure it in a few days. Of course I could not get its exact value in dollars and cents, but I could come so near it that the result would be satisfactory." Owing to the peculiar formation of the Comstock lode—clay and quartz-specked porphery—the Professor thought that the ground wire could not be used in Utah and Colorado.

Conversation then turned upon the heat in the lower levels of the Comstock mines. The thermometer in one mine runs up to 153°, and men find it almost impossible to work. "I have been thinking over it," said Mr. Edison, "and think the mines can be cooled without much difficulty." Here he drew a diagram, something like the following:

"There's your engine house," he said, putting his finger on A. Here's your shaft (B). And here's your levels," drawing the lines at C. "Now here," pointing to D, "the men are at work. When the thermometer stands at 103° say in the engine house (A), it's 153° where the men are working (D). They can't stand it. What makes

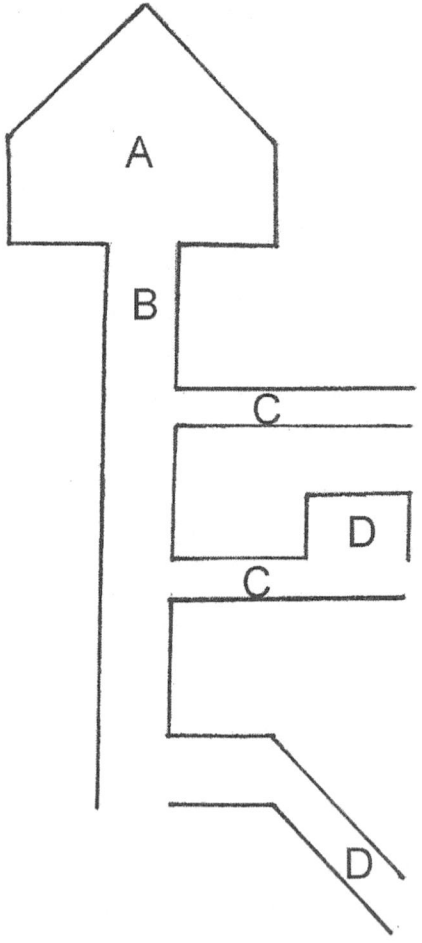

the heat? The water. That's what makes it. Pumps are at work all the time, trying to pump it out, but the walls and levels and all nooks are dripping with moisture. This water evaporates and generates more heat. The way to stop it is to stop the evaporation. Do you see that little pond out there," pointing to the little basin near the laboratory. "A pint of cotton-seed oil thrown upon it will stop the evaporation. All right. Now if they will take a few pails of cotton-seed oil and go down into the mine where the men are working (D), and awash the walls with the oil, covering the standing water, it ought to stop the evaporation in the mine. The generation of heat would cease, and the air become much cooler. I believe that if the thing was tried it would prove a success."

Here Stewart asked the Professor for a chew of tobacco. Mr. Edison handed a musty-looking plug, and a triangular lump went into the operator's mouth. He ground it once or twice, put on a wry face, and made a wild expectoration.

"Good Lord Tom," said he, "do you chew such stuff as that?"

The Professor was greatly amused. He bent over from the pillar and said, in a low tone, "I buy that, Stewart, to keep the boys from begging it from me."

But Stewart's remark had wiped the Comstock lode from his mind. His eye lighted on Fox, and he said, "Oh, how you can shoot around a corner! Shoot—I don't believe you could hit the top of that post yonder," pointing forty yards away. "Griff," he continued, turning to Griffin, "get the rifle and some cartridges and let us have a shoot."

Mr. Griffin produced a Winchester, and the Professor began to shove the cartridges into the side of the rifle. "I bought her out on the plains," he said, "for thirty-five dollars. She's cheap, but good." The party moved behind the factory, where a hill made the horizon a safe background, and the Professor's workmen came out to see the shooting. A glass insulator was placed on a gate post, thirty yards away. The rifle was sighted at 100 yards. Each man took a shot, and all missed until the turn of the Professor. He threw himself upon his knee as though at mass, took a long aim, and the insulator flew into the air in a thousand pieces. The schützenfest was kept up until there were no more cartridges.

The Professor has developed a passion for the rifle since his return from the West. A day or two ago he blazed away at the nob of a post and missed it. The ball whizzed between a barn and a Jerseyman, scaring the latter out of a day's growth and dumfounding the great inventor.

It was 6 o'clock. Col. Murphy and Mr. Stewart left for the train. Mr. Fox and the writer accepted an invitation to supper. The Professor lives in a neat house fronting the railroad, but a few steps from his laboratory. His mansion is substantially furnished. Everything is in perfect order, for a thrifty housewife holds the reins. After a bountiful meal, we returned to the laboratory. The Professor exhibited the drawings for a little machine that he had invented for the measurement of voltaic currents of electricity. One of his surprising discoveries is an ink that must prove a blessing to the blind. He threw a white powder into a bottle and poured a little water on it. After the bottle was well shaken he dipped a pen into the mixture, and wrote the word, "Boston" on different sheets of letter paper. The writing dried in a few minutes and raised itself on the paper the same as letters in books prepared expressly for the blind. The color of the writing was a clear white, and it looked as though it had been cut out and pasted upon the letter paper. It varied in height according to the sizing of the paper. If the surface was smooth and polished the letters were bold and clean cut. They were raised so much that a blind person accustomed to reading books with the fingers could easily decipher the manuscript. With the aid of the rack used by William Prescott in writing his "History of the Conquest of Mexico," and the use of Edison's preparation, blind men and women could correspond nearly as readily as other persons. The Professor says that he shall continue his experiments until he finds the paper best adapted to the use of the ink. Even while waiting for the ink to dry on one of the sheets, he made a singular discovery. He held the sheet to the blaze of a lamp. The word "Boston" turned a deep red, and finally became as black and shiny as ebony. On removing it from the blaze and looking at it on the surface, the ink changed to a color resembling the under side of a mushroom. The paper differed from that previously used, and the letters but slightly ridged the surface. The mixture acted upon the paper like an acid. It ate the impurities in the sheet, and left the word "Boston" as transparent as glass.

But a more surprising discovery was made. The ink seemed to act differently on different woods, and it is possible that Mr. Edison may find a wood on which the ink would stand out so much that artists could use it in drawing designs on blocks, and stereotyped plates could be taken without sending the blocks to a wood engraver. Such a discovery would almost put an end to wood engraving.

In answer to an inquiry concerning his ear trumpet, the Professor said he had received hundreds of letters from deaf persons. He had not yet perfected the machine, and could not say when he would do so. He needed such an instrument himself as much as anybody, for he is terribly deaf, and he should drive away at it as fast as possible.

His megaphone was not in the laboratory. This wonderful instrument has three funnels, modelled [sic] after a W. The two wings are funnels with tubing leading to each ear. The centre is a mouthpiece through with the operators talk. The instrument is simply an atmospherical telephone. With megaphones conversation can be carried on through a distance of from one to two miles and more.

"Do you see that house?" said Griffin during the afternoon. He pointed to a cottage more than a mile away. "Well they were talking with megaphones between the laboratory and that house the other day, and every whisper was heard." We were standing on the veranda, which commands a view so quietly beautiful that Mr. Edison says he saw none that surpassed it during his Western trip. "Take that clump of woods away off yonder beyond the grain stack, and you can hear every whisper that is uttered," said Griffin. "The instruments are pointed directly at each other like two cannons. If a bee flies between them you can hear his humming. A whisper is audible a thousand feet without using the speaking trumpet, and the noise made by walking through grass may be heard at a much greater distance."

The outside funnels are 6 feet 8 inches long and 27 ½ inches in diameter at the larger end. These funnels are each provided with a flexible ear tube, the end of which is placed in the ear. The speaking trumpet in the middle does not differ materially from the ordinary ones. It is a little longer and has a larger bell mouth.

When questioned concerning the noise on the Metropolitan Elevated Railroad, Mr. Edison said he believe he could greatly lessen it. He had not gone over a railroad bridge or culvert on his way to and from the Pacific coast without thinking of it. The Metropolitan road was one long bridge, and the noise was the same as that caused by the roaring of a train when rushing over a culvert or trestle. Mr. Edison thinks that the first cause of the noise is the roughness and looseness of the rails. He says that the rails on the elevated road are unnecessarily rough. They want planing down. Then they are loose on the sleepers, and the ends are too far apart. A temperature of 210° would not expand them enough to bring them together. As each car passes over the rails, they press down the end of one rail and strike the uplifted end of the next one. This causes the rattlecumbang so familiar to railroad travellers The sleepers are laid across the stringpieces, and this creates a roaring that might be obviated if the sleepers were mortised in the stringpieces, and the rails laid upon the same stringpieces. He was not prepared to give the public any further information at present, but he believed that the difficulty would be overcome before long. Within a few days he should resume his experiments on the road.

The Professor receives scores of letters daily, many of them being requests for money. The writers seem to think that he is coining money. He certainly does receive large royalties from his inventions, but his factory is expensive, and thousands are wasted in costly experiments. He appears to value money only as it assists him in his discoveries. A year or more ago he visited this city, and received some $30,000 in a lump. Half of this was immediately invested in a rural horse railroad, which has paid no interest on the investment, and the other half went for the most costly electrical instruments, chemicals, and other material for experiments. Before the sun went down the great inventor was obliged to borrow fifty cents to pay his fare homeward.

But all the letters are not monetary appeals. Months ago the Professor received a thousand foolscap pages of manuscript from a farmer's wife in the West, who wanted him to read it and write her what he thought of it.

It was a book on the theory of attraction. The Professor turned it over to his friend Griffin, who sent it back to the lady, and asked her to abridge it. The abridgment was received within a few weeks. The 1,000 pages had been reduced to 100. "And would you believe it," said Mr. Griffin, as he fondly fingered the pages during a conversation with the writer, "I read it clean through and found it an interesting and remarkable work. No man would believe that a farmer's wife wrote it. It shows an astonishing depth of research and great concentration of thought."

With the exception of the groove used for fastening the tinfoil on the cylinder, and screws to throw the steel point directly over the centre of the spiral cuttings surrounding the cylinder, there is no change in the speaking phonograph. A new clock-work instrument is nearly completed. No machines are sold, but any number can be hired on application to the Edison Speaking Phonograph Company on Broadway. The phonograph is yet in its infancy, but the day when it will be brought into general use cannot be far distant.

Some of the letters received are from spiritualists. They seriously claim that the great inventor is the especial favorite of spirits, and that his work is inspired by ghosts from the spirit land, who lavish all their gifts upon him, and unbeknown to him, direct his thoughts when on the eve of important discoveries. It is a singular fact that Mr. Edison's most wonderful discoveries have been made between midnight and daylight. He has not forgotten his long vigils in telegraph offices, and says his mind is clearer in the small hours of morning. He laughs at persons who speak of his genius. "Genius," he says, "is all bosh. Clean hard work is what does the business." His laboratory was recently visited by several French savans [sic], who urged him to visit Paris, and said that he would be received with royal honors. He declined on the ground that he had too much work to do. Those who know him well say that it is doubtful whether he would walk across the street to receive the Cross of the Legion of Honor.

While he was in the West Mr. Edison received a certificate of honorary membership from the Grand Circle of Workingmen of Mexico, who proudly claim him as a brother. When the writer left Menlo at 9 P.M. he was rolling the Spanish vowels under his tongue, and endeavoring to glean the purport of the parchment.

8. EDISON'S NEWEST MARVEL.[*]

SENDING CHEAP LIGHT, HEAT, AND POWER BY ELECTRICITY.

<small>ILLUMINATING GAS TO BE SUPERSEDED—EDISON SOLVING THE PROBLEM OF DIVIDING THE TOO GREAT BRILLIANCY FROM AN ELECTRIC MACHINE.</small>

Mr. Edison says that he has discovered how to make electricity a cheap and practicable substitute for illuminating gas. Many scientific men have worked assiduously in that direction, but with little success. A powerful electric light was the result of these experiments, but the problem of its division into many small lights was a puzzler. Gramme, Siemens, Brush, Wallace, and others produced at most ten lights from a single machine, but a single one of them was found to be impracticable for lighting aught save large foundries, mills, and workshops. It has been reserved for Mr. Edison to solve the difficult problem desired. This, he says, he has done within a few days. His experience with the telephone, however, has taught him to be cautious, and he is exerting himself to protect the new scientific marvel, which, he says, will make the use of gas for illumination a thing of the past.

Mr. Edison, besides his power of origination, has the faculty for developing the ideas and mechanical constructions of others.

* From the New York *Sun*, Monday September 16, 1878, p. 3. This article was widely reprinted in other newspapers across the country.

He visited the Roosevelt pianoforte factory in this city, and while examining the component parts of the instruments, made four suggestions so valuable that they have been patented. While in the mining district of the West, recently, he devised a means of determining the presence of gold below the surface without resorting to costly and laborious boring and blasting. While on a visit to William Wallace, the electrical machine manufacturer, in Ansonia, Conn., he was shown the lately perfected dynamo-electric machine for transmitting power by electricity. When power is applied to this machine, it will not only reproduce it, but will turn it into light. Although said by Edison to be more powerful than any other machine of the kind known, it will divide the light of the electricity produced into but ten separate lights. These being equal in power to 4,000 candles, their impractibility for general purposes is apparent. Each of these lights is in a substantial metal frame, capable of holding in a horizontal position two carbon plates, each twelve inches long, two and a half wide, and one-half thick. The upper and lower parts of the frame are insulated from each other, and one of the conducting wires is connected with each carbon. In the centre, and above the upper carbon, is an electro magnet in the circuit, with an armature, by means of which the upper carbon is separated from the lower as far as desired. Wires from the source of electricity are placed in the binding posts. The carbons being together, the circuit is closed, the electro magnet acts, raising and lowering the upper carbon enough to give a bright light. The light moves toward the opposite end from which it starts, then changes and goes back, always moving toward the place where the carbons are nearest together. If from any cause the light goes out, the circuit is broken and the electric magnet ceases to act. Instantly the upper magnet falls, the circuit is closed, it relights, and separates the carbon again.

Edison on returning home after his visit to Ansonia, studied and experimented with electric lights. On Friday last his efforts were crowned with success, and the project that has filled the minds of many scientific men for years was developed.

"I have it now!" he said, on Saturday, while vigorously turning the handle of a Ritchie inductive coil in his laboratory at Menlo Park, "and, singularly enough, I have obtained it through an entirely different process than that from which scientific men have ever sought to secure it. They have all been working in the same groove, and when it is known how I have accomplished my object, everybody will wonder why they have never thought of it, it is so simple. When ten lights have been produced by a single electric machine, it has been thought to be a great triumph of scientific skill. With the process I have just discovered, I can produce a thousand— aye, ten thousand—from one machine. Indeed, the number may be said to be infinite. When the brilliancy and cheapness of the lights are made known to the public—which will be in a few weeks, or just as soon as I can thoroughly protect the process—illumination by carbureted hydrogen gas will be discarded. With fifteen or twenty of these dynamo-electric machines recently perfected by Mr. Wallace, I can light the entire lower part of New York city, using a 500 horse power engine. I purpose to establish one of these light centres in Nassau street, whence wires can be run up town as far as the Cooper Institute, down to the Battery, and across to both rivers. These wires must be insulated, and laid in the ground in the same manner as gas pipes. I also propose to utilize the gas burners and chandeliers now in use. In each house I can place a light meter, whence those wires will pass through the house, tapping small metallic contrivances that may be placed over each burner. Then housekeepers may turn off their gas and send the meters back to the companies whence they came. Whenever it is desired to light a jet, it will only be necessary to touch a little spring near it. No matches are required.

"Again, the same wire that brings the light to you," Mr. Edison continued, "will also bring power and heat. With the power you can run an elevator, a sewing machine or any other mechanical contrivance that requires a motor, and by means of the heat you may cook your food. To utilize the heat, it will only be necessary to have the ovens or stoves properly arranged for its reception. This can be done at trifling cost. The dynamo-electric machine, called a telemachon, and which has already been described in THE SUN, may be run by water or steam power at a distance. When used in a large city the machine would of necessity be run by steam power. I have computed the relative cost of the light power and heat generated by the electricity transmitted to the telemachon to be but a fraction of the cost where obtained in the ordinary way. By a battery or steam power it is forty-six times cheaper, and by water power probably 95 per cent, cheaper."

It has been computed that by Edison's process the same amount of light that is given by 1,000 cubic feet of the carbureted hydrogen gas now used in this city, and for which from $2.50 to $3 is paid, may be obtained for from twelve to fifteen cents. Edison will soon give a public exhibition of his new invention.

The Sun.

9. EDISON'S ELECTRIC LIGHT.*

THE INVENTOR DECLARES IT PERFECT, AND MAKES AN EXHIBITION.

In the Laboratory at Menlo Park—How the Professor Discovered the Light—How he Decreases its Intensity and Utilizes the Surplus Electricity—How Public Buildings and Private Residences Can be Lighted.

The doors of the laboratory at Menlo Park are closed. No strangers are admitted. Intimate friends of Mr. Edison are barred out. The writer met the inventor in a falling rain on Friday afternoon. Mr. Edison was returning to his laboratory from dinner. His little daughter was dashing after him through the mud, hatless and without an umbrella. After some jocular fencing, the object of the visit was announced, and we went into the private office. Mr. Griffin, the Professor's Private Secretary, sat at a table, answering scores of letters received in the last mail. The Professor sat his dripping beaver upon the floor, and was about to light a clay pipe when a cigar was tendered. He accepted it with a twinkle of the eye, and the blue smoke began to curl above his head. Conversation had fairly opened, and the little daughter planted her elbows upon her father's knee, looked up into his face, and asked for money to buy candy. The father put his hand into his pocket, drew out a dime

---

* From the New York *Sun*, Sunday October 20, 1878, p. 6; clipping in Cummings scrapbook 4, NYPL.

and a cent, gave the little pleader the dime, and returned the cent to his pocket, after expressing a fear that so much candy might injure her teeth. The little one sped through the doorway like an arrow, and the inventor awaited further questions.

"Are you positive," I inquired, "that you have found a light that will take the place of gas and be much cheaper to consumers?"

"There can be no doubt about it," he replied.

"Is it an electric light?" I asked.

"It is," he answered. "Electricity, and nothing else."

When asked how he came to make his first experiments with a view to discovering such a light, he said that an electric light was no new discovery. He only claimed that he had found out how to utilize it. His first experiments were made long ago, when he had a laboratory in Newark, N. J. He had the usual trouble with the gas companies. They held him at their mercy, and did pretty much as they pleased. At one time they threatened to tear out his meter, cut off his gas, and throw him back upon tallow dips and kerosene. He spent a month trying to discover a light that would take the place of gas, but met with little success, and finally gave it up. After his removal to Menlo Park he made his own gas. Pointing to a sputtering burner, he smiled and said: "It was gas, but I can't say much for its quality. Griff here has always said that there was a mighty good chance for improvement, and at last we've got it, eh, Griff?"

Griff smiled and winked at the writer. Mr. Edison continued, "When I remember how the gas companies used to treat me, I must say that it gives me great pleasure to get square with them."

Here the Professor smoked a minute in silence. His mind was evidently running off to something else. A question drew him back to the electric light. He began to talk in a free and easy manner. "Last December," he said, "I experimented on the same line that I have struck now. I got together all the books that I could find, and read up on it, and thought I was left. From what I read I got an idea that my theory was utterly impracticable. Two months ago William Wallace invited me up to Ansonia. You know Wallace?"

"Well," said Mr. Edison, "he's a great brass manufacturer, and has a laboratory nearly perfect. He has all the different electric lights and the different machines for making them. Prof. Barker and Prof. Chandler were with me. I saw for the first time everything in practical operation. It was all before me. I saw the thing had not gone so far but that I had a chance. I saw that what had been done had never been made practically useful. The intense light had not been subdivided so that it could be brought into private houses. In all electric lights heretofore obtained the intensity of the light was great, and the quantity very low. I came back home, and made continuous experiments two nights in succession. I discovered the necessary secret, so simple that a bootblack might understand it. It suddenly came to me, the same as the secret of the speaking phonograph. It was real, and no phantom. I was as sure that it would work as I was that the phonograph would work. I made my first machine. It was a success. Since then I have made nearly a dozen machines, each different, and the last ones improvements upon those first made. The subdivision of the light is all right. The only thing to be accurately determined is its economy. I am already positive that it will be cheaper than gas, but have not yet determined how much cheaper. To determine its economy, I am now putting up a brick building back of my laboratory here. It is to be 125 feet long. I have already ordered two eighty-horse power engines for this building. I consider them the best engines in the country."

"What do you use the engines for?" I asked.

"To make the electricity," the Professor replied. "We use no batteries. It isn't necessary. We simply turn the power of steam into electricity, and the greater steam power we obtain the more electricity we get. One object in putting up this brick building is to ascertain how many electrical jets, each equal to one gas jet, can be obtained from a one-horse power."

"I've already told you," continued Mr. Edison, "that electric lights have had marked intensity, and a low quantity. I'm turning it the other way—reducing the intensity and increasing the quantity of light, as far as possible. It requires a good deal of experimenting to ascertain how far this can be done. You alter the nature of the electric light when it is done. I have already done it to a certain extent, and don't think that it was ever before attempted on the line on which I'm at work."

Here Mr. Edison dropped his cigar stump from his mouth, and turning to Griffin, asked for some chewing tobacco. The private secretary drew open a drawer and passed out a yellow cake as large as a dinner plate. The Professor tore away a chew, saying: "I'm partly indebted to THE SUN for this tobacco. It printed an article asserting that I chewed poor tobacco. That was so. The Lorillards saw the article, and sent me down a box of the best plug that ever went into a man's mouth. All the workmen have used it, and Griff says there is a marked moral improvement in the men. It seems, however, to have an opposite effect on Griff. You see he has salted away the last cake for his own use."

On being questioned concerning the articles of incorporation of the Edison Electric Light Company, recently filed with County Clerk Gumbleton of this city, Mr. Edison said that they proposed to light the city, public buildings, and private residences with electric lights. The electricity would be made by twenty or more engines, stationed in different parts of the city. Instead of manufacturing all the electricity at one central point, as gas companies make gas, there would be twenty stations. Each station would have an engine and several electric generating agencies. "You know," said the Professor, "that when electricity goes out it must always get back to where it went from. Therefore each station will have one grand return wire, with which separate wires will connect, thus forming the necessary electric circuit. I think that the engines will be powerful enough to furnish light to all houses within a circle of half a mile. We could lay the wires right through the gas pipes, and bring them into the houses. All that will be necessary will be to remove the gas burners and substitute electric burners. The light can be regulated by a screw the same as gas. You may have a bright light or not, as you wish. You can turn it down or up, just as you please, and can shut it off at any time. No match is needed to light it. You turn the cock, the electric connection is made, the platinum burner catches a proper degree of heat, and there is your light. There is neither blaze nor flame. There is no singing nor flickering. I don't pretend that it will give a much better light than gas, but it will be whiter and steadier than any known light. I do know now that it will be cheaper than gas. It will give no fumes nor smoke. No carbonic acid gas will be thrown off by combustion. It will be a great thing for compositors, engravers, and all forced to work during hot summer nights, for it will throw out hardly any heat. Shades may be used the same as shades upon gaslights, but there will be no real necessity for them. The wind can't blow it out. There can be no gas explosions, and no one will be suffocated because the electricity is turned on, for it cannot be turned on without lighting the burner. A person may have lamps made with flexible cords, and carry them from one point to another."

"Can you measure the amount of electricity used?" I asked.

"Well," Mr. Edison replied, "I have made no attempt to discover a meter. I know that it can be measured, but it may take some time to find out how. I propose that a man pay so much for so many burners, whether he uses them or not. If I find that this works an injustice, why I shall try to get up a meter, but I fear it will be very hard to do it."

Mr. Edison says that electric generating machines could be placed upon steamboats and locomotives, and the boats and cars lighted by the action of the engines, but the instant that the machinery stopped the lights would go out. He thinks that it may be necessary to have an extra engine in each station in cities, to be prepared for accidents. If the first engine should break down, the second one could be used to feed the lights. Country towns, with the use of the electric generating machines, could be lighted by water power. Any power could be used, provided it was strong enough to turn the shaft of the machine with the necessary rapidity.

"Where do you get the electricity to make your electric light?" was the question.

"From the power of the steam engine," he replied.

"How do you get it?" was the next inquiry.

"I want it so that one of your bootblacks can understand what you say."

Mr. Edison tore a fresh chew of tobacco from Griffin's cake, and was silent for a minute. "There are some things," he said, "that seem easily explained, and yet when you attempt an explanation you find it extremely difficult to make it. A boy asks you why you are able to talk, and what makes the sound. He may crowd you to a point that you could not explain. This steam power and electricity business is perfectly understood by scientific men. You say you never studied natural philosophy. I never studied it myself, and yet this steam-power electricity business is perfectly plain to me. I will try to explain it. If you beat a piece of iron with a hammer, the iron becomes hot. What makes it hot? Muscular power. The hammer and the iron is simply a machine for transferring muscular power into heat. You cannot waste power. If you use a certain amount of power to pull a weight into the air, what becomes of the power? It is in the weight, and in coming to the ground it will use the power that was exercised in raising it. You wind a clock. The power used in winding remains in the spring of the clock, and will be used as the clock runs down. Now here is a shaft. You slip a belt over the shaft, and it revolves by steam power. It doesn't take much power to turn the shaft. Now, lay a board upon the shaft, and stand upon the board. It requires extra steam power to turn the shaft. What becomes of this extra power? It goes into heat. How do you know it goes into heat? Because it sets the board that is pressing it on fire. Very well. Now, if you use this extra power to make electricity instead of heat, you make the electric light. As the amount of heat required to burn the board represents the extra power used by the steam engine, so the amount of electricity used to create the light represents the same power. Instead of a board upon a shaft or cylinder, you arrange magnets facing, but not touching each other. Their attraction has a tendency to stop the shaft the same as the board, and the power used to force the revolution generates electricity in these magnets instead of heat. So that your steam power is transformed into electricity, and by the use of simple mechanism this electricity is turned into light. If the conditions are such that the extra steam power can't go off into electricity or light, it will go off in heat, and vice versa."

The Professor here exhibited an electrical generating machine. It is what is known as a Wallace machine. A knot of magnets run around the cylinder facing each other. Wires were attached to it. Mr. Edison slipped a belt over the machine, and the engine used in his manufactory began to turn the cylinder. He touched the point of the wire on a small piece of metal near the window casing, and there was a flash of blinding white light. It was repeated at each touch. "There is your steam power turned into an electric light," he said.

"But how do you utilize the light?" was the next inquiry.

"Open your mouth," the Professor replied, with a pleasant twinkle in his eye. "I want to look at it."

The chaffing was too evident, and the mouth remained closed. Curiosity, however, led to the question as to what a man's mouth had to do with the utilizing of the light. Griffin began to smile, "I only wanted to see," said the Professor, "whether there was room enough in a reporter's cheek for nine rows of teeth."

It was a fair hit, but the rough edge was smoothed off by an exhibition of the light. It was a simple secret, but not one ready for publication. There was the light, clear, cold, and beautiful. The intense brightness was gone. There was nothing irritating to the eye. The mechanism was so simple and perfect that it explained itself. The strip of platinum that acted as burner did not burn. It was incandescent. It threw off a light pure and white. It was set in a gallows-like frame, but it glowed with the phosphorescent effulgence of the star Altaire. You could trace the veins in your hands and the spots and lines upon your finger nails by its brightness. All the surplus electricity had been turned off, and the platinum shone with a mellow radiance through the small glass globe that surrounded it. A turn of the screw, and its brightness became dazzling, or was reduced to the faintest glimmer of a glow-worm. It seemed perfect. The Professor gazed at it with pride.

"I would gladly give up the secret to the public," he said, "but the patents are not perfected. You know my trouble with the telephone in England. A burnt child dreads the fire. The public may not know, but I do know that if a description of this invention reaches Germany, Austria, and other countries in Europe before a patent is obtained, none can be secured. I lost the telephone patent in Germany through indulgence to the newspapers, but fortunately recovered it because the essential point in the invention was not fully described. I finally got the patent by the skin of my teeth, but at a greatly increased expense. Our patent laws are all right, but there is no knowing how long they may remain so if Congress keeps tinkering at them. The inventor who applies for a patent here secures it, whether a description of the invention has appeared in the newspapers or not. It is from no purely selfish motive that I keep my secret from the public. I have no wish to do it, but it is necessary for my own protection."

"How is this invention to affect the gas companies?" I asked.

"Oh, the gas companies," repeated Mr. Edison. "Well, of course, some of their plants will have to go. But it is not necessarily ruinous to them. All they have to do is to amend their charters, and take this in and run it. It ought not to hurt them much, as I can see. If the directors are wise, their stock ought not to depreciate to any great extent. The electric light, to be sure, is cheaper than gas, that is certain. If it is not as economical as I think, I shall make it so, for experiments convince me that there is plenty of margin. The gas companies can do away with their tanks, and slap engines into the stations necessary for the diffusion of the electric light. Their pipes can be used for the wires, and there would be quite a saving."

Mr. Edison said he hoped to have his invention in practical operation within six weeks. As soon as his engine house is built, he wants to place the light in every private residence in Menlo Park. He says he shall erect posts along the roads, and have a grand exhibition. "My object," he says, "is to keep the bugs out of the invention."

"What do you mean by keeping out the bugs?" I asked.

"Study its defects and make the thing complete before we move on the great cities," he replied.

There were two Wallace machines in the Professor's laboratory. One would produce light equal to 4,000 candles, and the other light equal to 24,000. Nevertheless he expressed an intention of constructing a machine of his own that would do its work and carry out his own ideas more satisfactorily.

Referring to the public impatience regarding the practicability and improvements on his inventions, he referred to his stock indicator, his electric pen, and his system of quadruplex messages as great triumphs. "Mr. Orton, in his annual report," the Professor added, "said that the quadruplex instrument saved the Western Union Company a half million dollars yearly. Well, now, I must say that the quadruplex instrument was infinitely harder to work to a success than this electric light. Just think of it, sending four messages at once over the same wire a thousand miles to Chicago or any other point, subject to all the conditions and changes of the weather! Why, if all the wires between here and Washington are down, and one of them is raised, that instant you practically have four wires to the capital.

"Take the speaking phonograph," continued Mr. Edison. "Some say it is a mere scientific toy of no practical use. I make improvements on it daily. It will take its place in the niche of public utility in good time. A sheet of tin foil that will hold 4,000 words is now placed in the machine automatically. A perfected instrument runs by machinery, and you can stop it at any time by pulling a cord. The register is perfect, and it does not get out of gear. While it does not talk as loud as some of the old machines, it catches the labials and consonants perfectly and throws them out more evenly. Any man can dictate to it at his leisure, and his office boy can run out a half dozen or dozen words, stop the machine by pulling the cord, and write out what is desired."

"No machine," said the Professor, "has ever been got up that did not require years for improvement. I ought to be allowed at least two years to improve mine. Now, old man, get out and let me go to work."

He dashed up stairs and went to work. The writer dashed to the depot and came to New York. Here he visited Henry A. Gumbleton's office and transcribed the following articles of incorporation:

State of New York, City and County of New York. ss—We, Tracy R. Edson, James H. Banker, Norvin Green, Robert L. Cutting, Jr., Grosvenor P. Lowrey, Robert M. Gallaway, Egisto P. Fabbri, George R. Kent, George W. Soren, and Charles F. Stone, all of the city of New York, in the county and State of New York, and Nathan G. Miller of Bridgeport, in the State of Connecticut, and Thomas A. Edison of Menlo Park, in the State of New Jersey, and George S. Hamlin of Rutherford Park, in the State of New Jersey, being desirous of forming a corporation pursuant to and in conformity with the act of the Legislature of the State of New York, passed Feb. 17, 1848, entitled "An act to authorize the formation of corporations for manufacturing, mining, mechanical, or chemical purposes," and the various acts of said Legislature additional thereto or amendatory thereof, have associated ourselves together for the purposes aforesaid, and in pursuance of the re-requirements of said acts, to make, sign, and acknowledge this certificate, and do hereby certify as follows:

First—The corporate name of the said company is "The Edison Electric Light Company."

Second—The objects for which the said company is formed are to own, manufacture, operate, and license the use of various apparatus used in producing light, heat, or power by electricity.

Third—The amount of the capital stock of the said company is three hundred thousand dollars.

Fourth—The number of shares of which the said company shall consist is three thousand.

Fifth—The term of the existence of said company is fifty years, from the fifteenth day of October, one thousand eight hundred and seventy-eight.

Sixth—The number of the trustees of the said company shall be thirteen, and the names of those who shall manage the concerns of the company for the first year are:

Tracy R. Edson, James H. Banker, Norvin Green, Robert L. Cutting, Jr., Grosvenor P. Lowrey, Robert M. Gallaway, Egisto P. Fabbri, George R. Kent, George W. Soren, Charles F. Stone, Nathan G. Miller, Thomas A. Edison, George S. Hamlin.

In witness whereof we have hereunto set our hands this 16th day of October, 1878.

THOMAS A. EDISON,E. P. FABBRI,

TRACY R. EDSON,GROSVENOR P. LOWREY,

JAMES H. BANKER,N. G. MILLER,

NORVIN GREEN,ROBERT M. GALLWAY,

ROBERT L. CUTTING, JR.,GEORGE W. SOREN,

G. R. KENT,CHAS. FRANCIS STONE,

GEORGE S. HAMLIN

10. THE NEW ELECTRIC LIGHT.*

EDISON IN GOOD SPIRITS, AND POSITIVE OF COMPLETE SUCCESS.

~

PREPARING FOR THE DECISIVE EXPERIMENT AT MENLO PARK—
THE SECRET TO BE KNOWN WHEN PATENTS ARE SECURED ON
THE CONTINENT—RESULT OF THE LAST EXPERIMENT.

A common cast iron stove heated Mr. Edison's office in Menlo Park yesterday. The day was cold, and the old stove sent out a cheerful heat. A small poker lay in an empty wood box behind it, and a rusty meat broiler hung on the wall. The telegraph instruments on the table near the front windows were uneasy. They kept up an incessant ticking. A row of boxed telephones was fastened to the compartment protecting the inner office. The following notice was posted near them:

PLEASE DO NOT TOUCH ANY
OF
THESE BOXES.
Menlo Park, Nov. 1, 1878—6 A.M.

The centre table and shelves were littered with odds and ends. Such books as the Circle of Sciences, Statistics of Mines and Mining, the Philadelphia Photographer, Chronological and Descriptive Index

---

* From the New York *Sun*, Friday November 15, 1878, p. 1; clipping in Cummings scrapbook 4, NYPL.

of Patents, the Electric Telegraph Apparatus, Knight's Mechanical Dictionary, the Journal of Telegraph Engineers, and Telegraph Law Cases, all newly bound, were conspicuous on the shelves. Evidently somebody had been trying to put things in order, but the Professor had been around and thrown everything into pi. Old ink bottles were scattered among the books, and a box of chalk or arsenic was half upset over a card bearing the words

DON'T TOUCH!

An open can of condensed milk sat upon three dusty scrap books, and was closely eyed by an exploded rocket. Newspapers in wrappers were covered with scraps of letter paper. Telephone handles hobnobbed with broken-down magnets, and a paper of fine-cut smoking tobacco was strewn over an oddly-constructed balance wheel. A gum-encrusted paste pot sat on the stove pipe hat worn by the Professor at the Mount Vincent reception, and a formidable butcher knife seemed ready to pitch into the pot from the shelf above. A picture of Thomas Alva Edison swung on the wall.

Griffin, Mr. Edison' private secretary, sat within the inner office writing letters. He looked up at the intruder, and recognized him with a smile.

"You won't find the body here," he said. You'll have to look nearer Philadelphia. You're the first reporter that's been here looking for that $25,000, but you're on the wrong track."

It was a grim humor, and as such grimly borne. In answer to inquiries Mr. Griffin said that "Tom was up stairs, and very busy." He hoped that he might not be disturbed. "He will see you," Griffin continued, "but don't bother him more than you can help."

Griffin sent for the Professor. I saw him coming down stairs through a glass partition. He was reading a letter and dropped mechanically from step to step. His hair was uncombed, and he had not been shaved for a week. His eyes were bright, but his face and hands were sooty. In one of his brown studies he had planted his elbow into some overturned acid, and there was a large red spot on the sleeve of his coat. The acid had nipped the flesh, for at intervals he rubbed his elbow with some vigor. A red silk handkerchief wrapped tight about his neck was a reminder of his recent attack of neuralgia. This was brought on by intense application. He said: "I sat eight hours within a foot of an electric light equal to the light of eight thousand candles fusing things in the electric arc. I wore glasses, of course, and completely forgot myself. When I stopped work, my face was burned as though it had been exposed to the hot sun. It was so sore that I couldn't wash it. Then the pains took me in the back of the neck and the head. I don't know whether it was neuralgia or something else. It makes no difference what they call it. I have had enough of it."

When asked whether he was ready to give the public the secret of his invention for the subdivision of the electric light, he said: "My patent in England is all right, but I have received no returns from the Continent. As soon as I am protected there, the whole thing will be made public. The invention is so simple that everybody can understand it."

"Are you still confident that you can produce an electric light that will be cheaper than gas and give as good a light?" I asked.

"I am more than ever confident," he replied. "I may say I am positive. I am experimenting night and day to ascertain the exact cost, but am already sure that the light will be much cheaper than gas. I have told you that I am putting up a brick building, 125 feet long, to practically demonstrate the utility and economy of my discovery. The building will be completed within two weeks. Come out and look at it."

The Professor covered his head with a new felt hat, and we went back of the laboratory. Brick walls had risen from the ground like magic. The building is almost ready for roofing. A hundred horse-power boiler is already in position, and the engine will be on the ground to-morrow. Mr. Edison engaged his engineer yesterday. It took but a moment. The man was familiar with the engine to be used, and came on from Boston and applied for a situation. The Professor read his letters of recommendation. "Be here as soon as you can," he said. "I'll give you $3 a day and extra pay for extra work."

With this eighty horse-power engine Mr. Edison purposes to light Menlo Park. He says he will start in with 2,000 lights. He will use telegraph poles, with arms across the tops, placing fifteen lights in each arm. The lights will be run across the plateau far out into the woods. "I shall also put them in all the houses around here," he said, "to see how the women folks handle them." The main object is to ascertain how many electric jets, each equal to one gas jet, can be obtained from a one-horse power. The experiment will also enable him to discover any defects and make the invention complete before trying to adapt it to the great cities.

"The thing cannot be done in a day," said Mr. Edison. "I am working on a principle without known laws to guide me, and the experiments demand both time and money. They require a thorough knowledge of chemistry, metallurgy, electricity, the laws of light and heat, steam engineering, magnetism, and other sciences. They must be made by a practical man, who is conversant with almost every science outside of bones and botany. I can't say that I am that man, but I do say that I have men with me who understand what I do not understand. The subdivision of light is perfectly successful, and its utility, to my mind, already assured."

No carbon is used in producing the light. Mr. Edison thinks that he will have it in practical operation at Menlo Park within six weeks or two months. "I see," he added, "in the morning's *Tribune* that I have given up my experiments with the electric light and severed my connection with the Metropolitan Elevated Railroad."

"Where did they get their information?" I asked.

After a grave pause he answered, "I give it up."

Ground has been broken for the erection of a private office to the left of the entrance to the laboratory. The brick walls are already going up, and the structure will soon be completed. It will cover a brick vault, in which valuable papers and unfinished inventions will be safe in case of fire.

After asserting the necessity for such a building, the Professor took a fresh chew of tobacco and went up stairs two steps at a time.

11. DAYLIGHT AT MIDNIGHT.*

MR EDISON'S LATEST EXPERIMENTS WITH THE ELECTRIC LIGHT.

—

No Fears about his Patents—Searching Astor Library—
He Thinks he Can Light and Heat Private Houses and Run
Sewing Machines with the Same Telegraphic Circuit.

Grimy and sooty, Thomas Edison yesterday descended from his Menlo Park laboratory. All his time and energy are now bent toward the development of the electric light. His eyes are bright and restless, and his motions quick and impulsive. It is easy to see that his thoughts are with his experiments when talking on the most common place subjects. "The thing is opening up before me," he says, "and I am daily making new and most important discoveries." He seems like a man walking in another world.

When asked concerning the report that an Examiner in the Patent Office had rejected his application for a patent upon a divisible electric light on the ground that it is an infringement upon an invention made by John W. Starr of Cincinnati in 1845, he said that it could not be true. "Prof. Morton, in an article printed in THE SUN last week, spoke of that invention, "he says. "The patent was taken out in England by a Mr. King, who was Starr's Patent Solicitor. It was for an electric lamp. The light came from incandescent carbon, and was identical with a light now on

---

* From the New York *Sun*, Friday November 22, 1878, p. 3; clipping in Cummings scrapbook 4, NYPL.

exhibition in New York by Sawyer & Mann."

Here Mr. Edison referred to Prof. Morton's article, which described the Starr light as consisting "of an airtight glass vessel, within which a small rod of platinum or carbon was so placed that it could be heated intensely by the passage of an electric current. The air was removed from within the vessel, in case a carbon rod was used, to prevent its combustion."

Mr. Edison says his invention is different. "I can't patent the divisibility of the electric light," he adds, "but I can patent the means that allows it. In other words, I can patent a lamp, or any device that will make this division. My application for a patent for a lamp is already before the Commissioner, and is taking its regular course. According to the rules of the Patent Office nothing concerning it can be divulged. I have heard that it is progressing favorably, and that is all I have heard. One thing is certain. My application does not conflict with Starr's invention. I have already received seven patents bearing on the electric light, and have filed three caveats. Five more similar applications are now under way. I have had a man in the Astor Library search the French and English patent records and scientific journals from the earliest dates down to the past fortnight, and nothing like my arrangements has been revealed.

"Scientific journals," says Mr. Edison, "frequently pick me up on misstatements by reporters. For instance, one of your attachés some time ago represented me as saying that I could produce 10,000 lights with one electric machine. He misunderstood me. I meant that I could produce 10,000 lights from one station, and there might be from thirty to fifty machines in each station. If you remember, I propose to light cities from electric stations, the wires covering so many blocks. There is a vast difference in the two statements.

"Now," continued Mr. Edison, "I find the scientific journals going for me on the idea of the supplying power. I fully understand

and know that there is an enormous loss in transforming steam power into electric power, but if I can sell the latter power for three times more than the original cost of the steam power, there is a good profit notwithstanding this loss. To illustrate: With the same wires used for the electric light I could put electric power into a private house that would run a sewing machine. Now the loss in furnishing this power might be seventy-five per cent, but if I could get ten cents a day for supplying the power to run each machine I should make an enormous profit."

"Do you mean to tell me that you can supply this power with the same engines, the same electric machines, and the same wires that you would use to produce the electric light?" I asked.

"I mean just that," answered the Professor, smiling, "and if I am not mistaken in the purport of my experiments, I mean to say that I can light and heat private houses and supply electric power for sewing or other machines with the same telegraphic circuit. I have made a discovery by which I think I can concentrate and use the electric heat which has heretofore been wasted. My discovery is a substance which, placed between two metal plates, holds the heat, and gives enough of it to cook beefsteak or make tea, soup, or coffee. I am not yet positive about it. It may prove a failure, but the result of my experiments justifies me in saying that it bids fair to be a success. We know that we can transform electricity into heat, light, and motive power. There is less loss in turning it into light and heat than in making it a motive power. The loss in the motive power is considerable, and is probably due to defects in the electric machine. The loss in transforming electricity into light and heat is inconsiderable."

Here Mr. Edison folded his arms and relapsed into a brown study. The electric light was recalled, and he resumed the conversation. "The success of the light," he said, "is assured. Why, if I'm not mistaken in my experiments thus far, I can go to New York and buy gas from the gas companies at the rate they are selling it at, and through the use of a gas engine turn it into motive power for the production of electricity, and transform the electricity into light, giving a better light, more of it, and fully as cheap, if not cheaper, than the original gaslight. This is no more wonderful than what was done by Prof. Apony with a German electro-machine, driven by a five-horse petroleum engine. I read it in THE SUN. The engine consumed a little over six and a half pounds of crude petroleum per hour. It produced a stream of electricity having an illuminating power three times greater than that of the petroleum used."

Mr. Edison reflected a moment, and then said: "Now, write this down as I give it to you. It is a law that I have figured out and verified. Take a gas jet having ten inches of radiating surface, burning five feet of gas an hour, giving a light equal to fifteen candles. If the radiating surface of this gas jet could by any possible means be reduced from ten inches to an eighth of an inch without losing any heat, the temperature of this surface would be enormously increased, and would give a light equal to that of a Joblochkoff electric candle, or a light equal to the light of 550 candles.

"In other words," continued Mr. Edison, "if it were possible to concentrate a gas jet to the size of a pea without losing its heat, we could get the light of 550 candles instead of fifteen. I've figured this out and verified it. This shows where electricity gains; for while it is not possible to concentrate the gas jet, keeping all the heat in it, you can turn it into electricity, and practically concentrate it by the use of a gas engine. If the gas men could concentrate their gas jets there would be no use for an electric light. They furnish ninety per cent of heat, which you don't want, to give you ten per cent of light. When you turn their gas into electricity you get ninety per cent of light and only ten per cent of heat."

Mr. Edison then retired to his laboratory. All visitors and inquirers concerning the electric light are now referred back to the office of the company in this city. The Professor will see only his intimate friends and very few of them. The preparations for the decisive experiment at Menlo Park are well under way. The engine house is roofed and the engine at the railroad depot. The wires will probably be up within six weeks and midnight at Menlo Park will be flooded with daylight.

12. THE NEW ELECTRIC LIGHTS.*

WHAT MR. EDISON THINKS OF THE RECENT ENGLISH INVENTION.

RIO JANEIRO FIRST TO USE THE LIGHT—MR. EDISON ON
WERDERMANN'S CARBON LIGHT—A WONDERFUL IMPROVEMENT
IN THE TELEPHONE—A MARVELLOUS TYPE-WRITER—A PATENT
SECURED THROUGH THE USE OF THE ATLANTIC CABLE.

A small American flag fluttered from the roof of the new engine
house in the rear of Mr. Edison's Menlo Park laboratory
on Saturday. The building is nearly completed. The boiler is in
position, but the engine remains on the platform cars near the
depot. A score or more of men are at work in the yard, and every
possible preparation is being made for the decisive experiment with
the electric light. The inventor's faith in the practicability of his
invention is shown by the estimated cost of the experiment. He
purposes to start in at Menlo with 2,000 lights, using telegraph poles
with fifteen lights on each arm. This experiment, including the cost
of the buildings, engine, generating machines, and everything, he
thinks will eat up from $75,000 to $100,000. One of his assistants
says: "I always go on the principle of adding thirty per cent to the
estimated cost of all experiments, and find myself nearer right in
the end. It will cost nearer $125,000 than $100,000 to make this
experiment at Menlo."

---

* In New York *Sun* on Monday November 25, 1878, p. 1; clipping in Cummings scrapbook 4, NYPL.

Mr. Edison nodded his head approvingly. "Whatever the cost," he said, "it will prove the practicability, cheapness, and utility of this subdivision of the electric light beyond a doubt."

The downfall in gas stocks has been followed by a corresponding increase in the price of shares of the Edison Company. It is understood that 260 has been offered and accepted. The men

at Menlo say that they are told that the stock of a prominent gas company was recently offered by auction at Nicolay's in Wall street. The highest bid was seventy. The shares were withdrawn.

On receiving this information the writer said that a well-informed Wall street gentleman thought the purchase of gas stocks was the best investment that a moneyed man could make at this time. He thought they had touched bottom, and would rapidly come to the surface under the influence exerted by the printed opinions of Prof. Morton and other savans [sic]. Morton's articles had excited grave doubts as to the utility of Mr. Edison's invention. But even supposing that the Edison idea was practicable, there was always the chance that the inventor might die and his invention relapse into obscurity for want of development.

Mr. Edison tore a soiled silk handkerchief from his neck, and laughed outright.

"Has this Wall street gentleman any money?" asked one of his assistants.

"I don't know," was the answer. "Why do you ask?"

"Because if he has any money, and backs up his opinion with it, he's sure to lose it," he said.

THE RECORDS OF THE EXPERIMENTS.

Mr. Edison broke into the conversation. He seemed to think that Prof. Morton owned gas stock, or was writing in the interest of the gas companies, although he gave no expression of such an opinion. His assistants, however, freely uttered their suspicions. "My death," said Mr. Edison, "could not destroy the utility of my invention. The thing is proved to be of practical use. I keep a record of each experiment, and file away every drawing. Not an etching or scrap of paper is destroyed. Batch," turning to one of his assistants, who was figuring out a problem, "show him the records of our experiments with the electric light."

Batchelor disappeared and returned staggering under a load of books and foolscap paper. They would have made the eyes of an old-paper man dance with delight. There were over a thousand sheets of drawings alone. A notary's seal was stamped on each sheet. A stranger might have taken them for a manuscript edition of Euclid sandwiched between scores of algebraic equations. The record was complete. There was not a lapse. "There they are," continued Mr. Edison. "Look them over for yourself. Before we get through with our experiments on this light, we shall have a pile of papers that will reach from the floor to the ceiling."

Confident himself in the success of his invention, Mr. Edison seems pleased with indications of public confidence. The well-known firm of Fabbri & Chauncey have bought from the Edison Company the exclusive right to the use of the new electric light in South America. From one of the officers of the steamship City of Rio Janeiro the writer learns that the contract for lighting the Brazilian capital expires very soon. In view of the franchise bought by Fabbri & Chauncey, he thought it not improbable that Rio Janeiro might be the first city in the world lighted by electricity.

THE RECENT ENGLISH INVENTION.

In further conversation I alluded to reports that gentlemen in England and France had discovered means for the subdivision of electric light. I drew from my pocket a slip from the London *News* announcing that England, and not America, had solved the great problem of the divisibility of this light. A Mr. Werdermann had given a public exhibition of an invention for lighting with a divided electric current. The current was generated by means of a two-horse power Gramme's plating machine and was conducted along a cable serving to light two lamps, stated to have the illuminating power of about 360 candles each. The light, which was perfectly steady, and in the room was soft and sunlike, could be looked at without discomfort, though it was not shaded. The next thing, the larger lamps being extinguished, was the exhibition of ten smaller lamps, fed by the same current. From the cable, which might be said to represent a gas main, a wire, answering to a service pipe, ascended to the positive electrode of each light. The construction of the lamp may be thus described: An upright rod of carbon (the positive electrode), resembling an ordinary slate pencil, touching the centre of the under side of a disk of carbon (the negative electrode), which was somewhat like a half pound weight. Briefly the lamp may be roughly compared to a common weight balanced on a sharpened slate pencil. A balance suspended from the apparatus served to keep the positive electrode in its place as the carbon would be consumed. Another cable and a special wire united all the negative electrodes, and at the last lamp the current went to earth. The connection having been made, the ten lamps were at once lighted; each light was stated to be of about forty-candle power. The lamps burned steadily, with a beautifully soft and clear white light. First one of the ten lights was then extinguished, and afterward a second, the only effect on the remainder being that they became slightly more brilliant, as gas will sometimes be under similar circumstances. Mr. Werdermann explained that this would not really be the case, as there would be an arrangement by which, on the extinction of a light, the current would be directed along a supplementary wire, equal in resistance to the consumption of the light, so that the resistance in the case of the other lights on the circuit would remain unaffected.

Mr. Werdermann, with the aid of diagrams, explained his invention. It was well known, he said, that when they burned an electric light of two carbons, one carbon burned out in a crater, while the other burned to a point; thus the one carbon was consumed twice as fast as the other. It occurred to him to find out what change would take place if he varied the section of the electrodes— that was, if he made one carbon smaller and the other larger. He then discovered this curious fact, that when he had an electric arc of one inch he could not maintain that arc if he diminished the section of his positive carbon. He went on reducing the section or magnitude of his positive carbon, the electric arc becoming lessened in regular proportion, until the difference of section of the two electrodes was as one to sixty-four and then the electric arc was so far reduced as that the two carbons were in contact. As the section of the positive carbon was diminished, the glowing portion of that carbon increased in length, while the heat imparted to the negative carbon was reduced. In effect, when the sections of carbon had reached the relative proportions just stated, and were in contact, the electric arc was infinitesimally small, the negative electrode was not consumed, while the positive electrode was incandescent. Light was therefore not only supplied by the electric arc, but was also furnished by the incandescent carbon of the positive electrode. It then only required a simple mechanical arrangement to keep the positive pole, as it consumed, in regular contact with the negative pole, and the difficulty which had hitherto stood in the way of using a number of lights from one current was overcome.

"Have you seen any account of this invention?" I asked.

Mr. Edison smiled, and laid before me an English weekly

newspaper, containing a fully illustrated description of the Werdermann subdivision. "For your information," he said, which a quiet chuckle. It was carefully read, and did not vary materially from the account given above.

MR. EDISON ON THE WERDERMANN LIGHT.

Mr. Edison then sad: "This Werdermann light neither forestalls nor conflicts with my invention. I use no carbon. My light does not burn itself out. The Werdermann light, however, is a good thing. It allows more subdivision than the Joblochkoff; but the maintenance of attention and consumption of carbon cost more than the horse power used to keep it running. It doesn't allow of a reliable subdivision."

"Why doesn't it?" I asked.

"Because," said Mr. Edison, "you must keep fooling with the lamp all the time to keep it going. To be of utility, the subdivision must be absolutely constant. As a carbon lamp, I like it very much. But the carbon lamp won't do. I have made repeated experiments with carbon, and at one time made a lamp very similar to this one of Werdermann's."

"Then you think Werdermann's invention of no practical utility?"

"No, I won't say that," he replied. "It may come into use in many places, but it can never become of general use. The lamp for the people must be so simple in its construction that any fool or mule can use it. I am making such a lamp. You turn your faucet, and there is your light. You turn your faucet, and your light is gone. There is no dickering and no feeding. Your light is there when you want it, and when you don't want it it disappears. It is so simple in its design that a child can understand it and use it, and it costs no more than the horse power used to turn the generating machine.

The current is noiselessly carried from one light to the other, and there is no rush or frying-pan sizzle as is the case with the carbon points. And then there is the question of supply. With a four-cell battery I can get more light with my invention than any carbon light so far devised, with the same amount of electric force, and I can see nothing to prevent its application to stronger currents on a larger scale."

PATENTS BY CABLE.

The loud ticking of a telegraph instrument interrupted the conversation. We were seated in the little office off the main entrance to the laboratory. Mr. Edison ran to a telephone boxed on the wall. I could hear a womanly voice informing him that dinner was ready. He resides an eighth of a mile from the laboratory. "All right," he responded, and resumed the conversation.

"These fellows across the water are working the light for certain," he said. "I see that seven patents were granted in England alone between Oct. 29 and Nov. 4. About one-tenth of all the patents now taken out in Great Britain are for electric lights. I patent all the discoveries made in my experiments both here and in England. And sometimes we make quick work of it. Last week I made a discovery at 4 o'clock in the afternoon. I got a wire from here to Plainfield, where my solicitor lives, and brought him into the telegraph office at that place. I wired him my discovery. He drew up the specifications on the spot, and about 9 o'clock that night cabled an application for a patent to London. Before I was out of bed the next morning I received word from London that my application had been filed in the English Patent Office. The application was filed at noon, and I received my information about 7 in the morning, five hours before the filing. The difference between London and New York time explains the thing."

NONSENSE AND COMMON SENSE.

Mr. Edison says that the cost of the manœuvre did not exceed eighty dollars. He thought that the total cost of his electric light patents might amount to $17,000. A minute afterward he picked up the slip from the London *News*, and quickly became interested. His eyes lighted with humor, and a smirk encircled his mouth. "Now, look here," said he, reading from the slip:

Mr. Werdermann stated that he did not believe in the principle of the indefinite division of a current of equal strength, which Mr. Edison had been said to affirm.

"Well," he drawled, "as Lord John Russell said, the sources of the information obtained by her Majesty's Government are quite incomprehensible. I don't and never did affirm any such nonsense as that. It is nonsense on its face." He read further:

In his (Mr. Werdermann's) opinion a current equal to maintaining a hundred lights would not maintain five hundred lights any more than an amount of gas only sufficient for a hundred burners would supply five hundred burners. But, he added, the proportion of horse power absorbed by each light would be less as the number of lights was greater. The greater the number of lights the greater would be the quantity of the current required, and the longer the distance traversed by the current the larger must be the electro-motive power.

"That's common sense," added Mr. Edison. "No sane man can doubt it."

"I understand," I said, "that you propose to feed a certain number of lights from one current of electricity. Now, when that current strikes the first light, and what is not wanted is carried on to the next lights, would it not be a great deal stronger than when it reached the last light? Might it not be so strong as to melt the contrivance used to send it on if it did not melt the burner?"

"You can only carry so much water through a two-inch pipe," he answered. "A certain amount of gas will go through a gas main, and no more. A wire will take only so much electricity. We arrange the conductor in such a manner that the current at all points is the same, and that is all there is of it."

SIX LIGHTS TO ONE HORSE POWER.

In further conversation, Mr. Edison said he was getting four times as much light with the same force as he did when he first began his experiments. He found the generating machines in use faulty, and was making two improved machines of his own, with the view of turning the greatest amount of horse power into electricity with the least possible loss. These machines are to be especially applicable to his electric light. From his experiments he was satisfied that he could get three lights equal to an ordinary gas jet for each horse power, with each old generating machine, and six with the new machines. The writer then called the inventor's attention to the following extract from the London correspondence of the Liverpool *Post*:

Some important information has reached me on the nature of Edison's electric light. It is formed by a coil of platinum being placed over a wire heated by the electric current. The coil itself is the source of light, the current sent through it being strong enough to make it white hot and self-luminous. The difficulty to be overcome at this point of the invention was the liability of the wire to fuse and spoil the light—a difficulty which Mr. Edison claims to have obviated by the introduction of a simple device which, by the expansion of a small bar the instant the heat of the coil approaches the fusing point of the platinum, interposes as a check to the flow of the current through the coil. This automatic arrangement secures, it is said, an even flow of electricity through the coil, and consequently a steady flow of pure light.

"This was evidently written," said Mr. Edison, "by some one who had seen one of my earliest experiments. It does not tally at all with my latest light. There is no bar, and the whole thing is entirely different. I fancy that I can soon give a complete explanation to the public. Batch., where is the lamp that this article evidently refers to?"

Mr. Batchelor hunted it up. It was found among the cast-off experiments, broken and dusty, and was exhibited amid the laughter of Mr. Edison's assistants.

The telegraphic instrument again began to tick, and the sooty savan [sic] jumped to the telephone. The womanly voice repeated that dinner was ready. "Never mind," shouted Mr. Edison in response. "Can't you send up a lunch for three?"

ENGLAND AWAKE.

He received a faint answer, but owing to his deafness did not hear it. "hallo, hallo!" he shouted. The womanly voice answered. "Send up a lunch for three!" he cried, and resumed his seat. Mr. Batchelor handed him a slip from the London *Times*. He passed it to the writer, saying: "Do you remember the controversy in England concerning the merits of the carbon telephone? Hughes and Priest of the British Postal Office both condemned it. Hughes said it didn't work at all. Well, now read what the British Thunderer says about it."

The extract was cut from the London *Times* of Nov. 12. It gives an account of "some interesting and valuable experiments on electric telephony, between Norwich and London, under the most adverse circumstances of bad weather and powerful induction of neighboring wires. These experiments," the *Times* says, "had for their object the confirmation of some extraordinary statements that had appeared in the American journals respecting the ability of the carbon telephone, one of Mr. Edison's numerous inventions, and perhaps one of the most important, to work over great distances and under conditions favorable to other systems—conditions which up to the present time have been the chief obstacle in the practical use of electric telephony." The wire stretched from Messrs. Colman's works in Norwich to their office in London, or a little over 115 miles. It ran over the same poles as other wires. When the experiments began, the incessant crackling and bubbling sounds in the receivers revealed the fact that the adjoining wires were being worked to their fullest capacity, and that induction could hardly be worse. The first exclamation uttered in Norwich was heard perfectly in London. Conversation was carried on without difficulty, and the Yankee accent of Mr. Edison's agent was distinctly recognizable. Remarks passed on the weather showed that there was a storm of sleet at both ends. The conversation was best heard when carried on a little below the ordinary tone of voice. Toward 9 P.M. the induction disturbances grew less, but were still considerable. The voices from Norwich were louder, and the individuality of the speakers more marked.

ASTOUNDING DISCOVERIES.

Commenting on these experiments, the *Times* says:

Remarkable as they were, they appear to be outstripped by what has been achieved in America through the same instruments. Mr. Prescott, the chief electrician of the Western Union Telegraph Company, says that they have been successfully used when included in a Morse circuit, and that several stations could exchange business telephonically upon a circuit that was being worked quadruplex without disturbing the latter.

This statement was so astounding that I asked Mr. Edison if it was true.

"It was done by Henry Bentley of Philadelphia," he replied. "It has also been tried with success over a wire 720 miles long." The *Times* continues:

Mr. Edison has lately made a new and improved receiver to the instrument, of which he says in a recent letter to Col. Goureaud that by its means Batchelor, one of Mr. Edison's assistants, "heard a whisper last night fifteen feet away from the receiver, and ordinary conversation comes out into the room almost as originally spoken." If this receiver proves as practicable as the carbon transmitter, a new era has opened in electric telephony, and soon we may hope to have the speeches in the House of Commons heard at all the clubs in the metropolis.

This statement was even more astounding than the other. Mr. Edison says it is strictly true. He has one of these receivers now at work in his laboratory. Words spoken miles away are uttered within the room, and so plainly that they are distinctly heard in any part of it. A whisper is heard from fifteen to twenty feet. The great inventor says: "I think it entirely practicable for an audience miles away to hear every word uttered by Henry Ward Beecher in delivering his sermons at Plymouth Church. The debates in Congress, as well as those in the House of Commons, may also be heard by an audience anywhere."

ANOTHER REMARKABLE INVENTION.

One of the most remarkable of Mr. Edison's inventions is a stencil type-writer. He patented it in 1876, but threw it aside without improving it to begin the experiments which resulted in the discovery of the telephone. To use his own words: "I chucked it into a dark closet, after getting into the telephone business, and there it remained until the other day, when my old friend J.U. Mackenzie came in with a similar idea in his head. I yanked the thing out of the closet, turned it over to him, and said: You want something to do. Work that up. He took it and cleared out. To-day he came back, and instead of dropping the thing, as I expected, he seems to have made a perfect success of it."

Mr. Edison's idea seems to have been founded on the use of needle-pointed type, but Mr. Mackenzie gives them chisel points, which is more durable and makes the stencil fully as distinct. These types are placed in a type-writer, worked with the fingers like a pianoforte. Each written page may be placed upon a Gordon press or an electric pen press and as many copies worked off as wanted. When mistakes are made the type may be corrected, and the lines may be justified or spaced out, as desired. It must prove of great value to lawyers, as they might make as many copies of their briefs or other documents as they required. By its use any man might write and print his own circulars. Small daily country newspapers might also be printed. Crude as the invention necessarily is, Mr. Mackenzie printed 148 copies of a single page on Saturday in thirty minutes. He says his chisel-pointed letters are subject to modification, but he regards the success of the stencil type writer as assured, and there is undoubtedly a fortune in store for its inventors.

THE LATEST DISCOVERY.

After an explanation of this invention, Mr. Edison became jocose. Story followed story, and there were frequent sallies of wit and humor. As we were drawing on our overcoats preparatory to departure the great inventor said: "I am now about to tell you something that will astonish all electricians. I am prepared to send a current of electricity from here to Philadelphia without any wire."

"Why Al, (his second name is Alva, and many of his friends call him Al) that's impossible," said Mackenzie, who is an old telegraph operator.

"Oh, no," answered Mr. Edison. "It can be done, and I know it. It is the result of a recent discovery."

"How?" inquired Mackenzie.

"Store it up in a condenser and send it there by express," was the reply. "Now don't give it away to any of these newspaper men."

The Sun.

13. THE DARKNESS DISPELLED.*

AN ESTIMATE OF THE CHEAPNESS OF EDISON'S ELECTRIC LIGHT.

SIX HUNDRED LIGHTS, EACH EQUAL IN INTENSITY TO A COMMON GAS JET, FROM A HUNDRED-HORSE-POWER ENGINE—WHAT IS TO BE THE COST OF LIGHTING STEWART'S STORE, BOOTH'S THEATRE, AND THE FIFTH AVENUE HOTEL.

His subdivision of the electric light being perfect and its utility an accomplished fact, Mr. Edison has turned his attention toward a new generating machine. He has already invented a meter for the light, which works to a charm. With the Wallace generating machine, Mr. Edison says he can get no more than three electric jets, each equal to one gas jet, from a one-horse-power engine. Five months ago he began experimenting with a view to saving the loss of electricity by the Wallace and other machines. The perfection of his apparatus for subdivision interrupted these experiments, but now he is giving almost his sole attention to them. Two weeks ago he announced that he had doubled the productive power of the Wallace machine. He felt sure that he could turn out six lights to a one-horse power. With his new eighty-horse-power engine, he could force out 480 electric lights, in the place of the 240 generated by the Wallace machine.

---

* From the New York *Sun*, Saturday December 14, 1878, p. 1; clipping in Cummings scrapbook 4, NYPL.

On Wednesday at 10 A.M. he began a new series of experiments, believing that he could still further increase the power of his own generator. The experiments lasted all night and until 10 A.M. of Thursday. They were made in the presence of a well-known New York lawyer, and were successful. More light was obtained, but exactly how much more was not ascertained. The inventor went to bed at 11 A.M. thoroughly exhausted, and slept until daylight yesterday morning. He visited the city yesterday, and returned to Menlo last night.

The announcement that 480 electric lights, equal to common gaslights, can be produced by the use of his proposed generator and an eighty-horse-power engine gives us data by which we may ascertain the cost of the light with a tolerable degree of accuracy. There are ninety gas burners in THE SUN composing room. They are used, say, twelve hours out of the twenty-four. They cost us last week $54, or $7.71 a night, a little over 8½ cents apiece. Six hundred such lights, burned for the same period, would cost a little over $51 per night. At an over-liberal estimate, it would take two tons of coal to run an engine of 100-horse power twenty-four hours. This would cost $7. The services of two engineers would cost $6 more, or $13 in all. For this $13, then, if Mr. Edison's statement is correct, we could get 600 lights, and burn them twenty-four hours. For $6.50 we could burn them twelve hours, leaving the figures thus:

600 gas jets..	$51.00
600 electric jets	6.50
Difference in favor of the latter	$44.50

Take another estimate. Booth's Theatre has 1,500 gas burners. It costs from $30 to $35 per night to burn them three hours and a half. If Mr. Edison's estimate is correct, the managers could burn the same number of electric lights, of equal brilliancy, twenty-four hours for $32.50, or about $4.60 for the three hours and a half.

The figures would stand thus:

1,500 gas jets	$30.00
1,500 electric lights	<u>4.60</u>
Difference in favor of the latter	$25.40

There are 3,500 gas burners in A.T. Stewart's retail store. It is difficult to ascertain an average of the time they are burned. They cost, however, over $50 per night. Carrying out the Booth's Theatre estimate, Judge Hilton and his partners would save over $40 a night by the use of the electric lights.

The Fifth Avenue Hotel has 2,050 burners. They burn 500,000 feet of gas a month in the best times. At $2 per thousand this would amount to about $34 per night. By the use of the electric light they would save nearly $27 per night, estimating on the Booth's Theatre ratio.

All these figures are based on Edison's estimate of the power of his generator. This generator has not yet been made. The inventor bases his assertion on the result of his experiments alone. He does, however, positively assert that he can get three lights per horse power, each equal to a gas jet, through the use of a Wallace machine. This is solid data. The saving per night in the different buildings above mentioned, the conditions being the same, would be as follows if the Wallace machines were used:

*SUN office	$22.25
Booth's Theatre	12.70
A.T. Stewart's	20.00
Fifth Avenue Hotel	14.25

*Allowing 600 lights to burn twelve hours.

The above figures are given for what they are worth. Edison positively asserts that he can get three lights, each equal to common gas jets, to a one-horse power by the use of a Wallace machine, and more than double that number by the use of his proposed generator. Any school-boy can ascertain the cost of coal and attendance on an engine and work out the problem for himself. A hundred-horse-power engine would cost $2,500, and a generating machine as much more, so that the interest on the investment would amount to a small sum. The cost of the burners is not given, but they are expensive, as platinum is used. Once set, however, there would be no further expense.

14. THE GENIE OF MENLO PARK.*

AN EFFORT TO FATHOM THE MYSTERY OF THE NEW ELECTRIC LIGHT.

WHY THE CHEVALIER EDISON DOES NOT TAKE OUT HIS PATENTS—THE TROUBLE IN ENGLAND—EDISON'S OWN ESTIMATE OF THE CHEAPNESS OF THE NEW LIGHT—THE SPEAKING PHONOGRAPH.

When one of Mr. Edison's newspaper friends asked on Tuesday to see the specifications of all patents relating to the electric light, he was greeted with a hearty laugh. Then Chevalier Edison himself examined his cheek, and Private Secretary Griffin called for a Winchester rifle. "Really," said Griffin, "this request is the essence of newspaper audacity. You deserve shooting."

"But," said the writer, turning to the Chevalier of the Legion of Honor, "I suppose that a copy of the specifications in the patents granted might be obtained by application at the Patent Office, I thought you might be willing to save me the trip to Washington."

"Oh, no," said the Chevalier, unbuttoning his acid-eaten coat. "You can't get the specifications at the Patent Office. They are kept secret until the patent is taken out. Several patents on the light have been granted, but I am not compelled to take them out inside of six months. The publication of these specifications at this time would

---

* From the New York *Sun*, Saturday December 14, 1878, p. 3; clipping in Cummings scrapbook 4, NYPL.

prevent my getting my patents in Europe. Otherwise, I would gladly give you the specifications. Read that."

The Chevalier threw down the following letter:

New York, Dec. 16, 1878.

THOMAS A. EDISON—*Dear Sir:* The newspaper men are doing you more harm, and producing more trouble. I am just in receipt of another opposition to your German patent on the speaking phonograph, which is based on the publication in the *Scientific American* reproduced in France.

As I told you at the time, I was fearful that this publication in the *Scientific American* would give trouble, and so it has turned out. I have written to my agents there, giving them the best instructions as to the manner of meeting this opposition.

Yours truly, L. W. Serrell,
Solicitor of Patents, 76 Chambers street.

"Of course," said the Chevalier, "I know that none of the newspaper boys would willingly injure me; yet that speaks for itself. The article in the *Scientific American*, having been translated into a French paper, found its way into Germany. The Germans caught the idea, and for months have been applying the principle to toys and other things. If my application is favorably received, it will do away with all their little arrangements, and so they have consolidated to oppose me throughout."

"By the way," continued the Chevalier, "you have sent me several letters inquiring the price of the speaking phonograph. The machines are now on the market. I have sold out my right to the Edison Phonograph Company of 66 Reade street. They are selling machines at from $100 to $125 each. They are also getting out a small machine that will be retailed at $5. It will talk as clearly as the most expensive machine, but will not receive so many words at a time."

The Chevalier's attention was called to a London despatch announcing that Arnaud, under cover of one Russell, is now contesting Edison's application for a patent in England. "Who is Arnaud?" we asked.

"He's a Frenchman," the Chevalier replied. "I have heard of him, and that is all. I don't know whether he lives in Paris or Bordeaux. He seems to have objected to my application blindly, without evidence on which to base an objection. And right here is where you newspaper men come in again," the Chevalier added, with a smile. "Read that."

He threw down a second letter:

LONDON, W.C., 4th Dec., 1878.

DEAR SIR: We very much regret to inform you that the application for Patent Electric Light No. 4,226, 1878, is opposed, and the grounds of objection are that Mr. Edison is not the first and true inventor of the whole or parts. We have not yet received the official notice, but hope to do so to-morrow. We have already taken steps in the case to resist the objection, and we have sent notice to Messrs. Herbert & Co., who are the agents of the objector, a Mr. Russell, but who the latter gentleman is we have not yet been able to ascertain. He does not appear to have any patent proceeding in connection with the electric light. We fully anticipate the objection is made vexatiously.

You will now fully appreciate our reason for wishing the invention kept secret until the opposition stage has passed. Yours truly, BREWER SPENSER.

The following is an official copy of Russell's protest:

[Patent Law Amended Act, 1852.]

I, John Huddart Russell, of 12 Cork street, in the County of Middlesex, hereby give notice that I object to the grant of letters patent to Thomas Alva Edison of Menlo Park, in the State of New Jersey, United States of America, for the invention of "Improvements in the method of and means for developing electric currents and lighting by electricity," as set forth in his petition recorded in the office of the Commissioners of Patents for Inventions on the 25th day of October, 1878, and my objection to the grant of such letters patent is as follows:

That the said Thomas Alva Edison is not the true and first inventor of the said improvements, or some of them.

Dated this 3d day of December, 1878.
HERBERT & Co., Agents.

"Now," said the Chevalier, "this paper was filed on Dec. 3. The case was privately argued before the Solicitor-General on the 16th. If Russell or Arnaud had any real grounds for objection, it seems to me that was the time to present them. They lodged no evidence to sustain the protest. Russell paid all costs, and secured an adjournment on the ground that he wanted time to get evidence. His protest is made in the dark. He doesn't know the secret of my light, and that's what's the matter."

Questioned concerning the perfection of his electric light, the Chevalier Edison said: "I am all right on my lamp. I don't care anything more about it. Every bit of heat is utilized to produce light as far as art will allow. The theoretical and practical results are perfectly satisfactory. My point now is the generator. The Wallace machine gives me three lights, each equal to a gas light, to a one-horse power. I feel sure that I can get six with an improved machine. Probably I can get more. Now, to make my grand practical experiment here in lighting Menlo Park, I should have to use twenty or thirty Wallace machines. They would cost me from $30,000 to $40,000. They would be useless afterward, for I know that I can make a generator of double their power. So I shall postpone the experiment until I find the machine that will give the greatest amount of electricity per horse power. I am making many little generators of different forms. I propose to run them with weights, figured down to the fractions of a horse power, and shall take the machine that generates the most electricity."

"If you can run a generator with weights," I said, "what is there to prevent a man from buying a generator and lighting his own house?"

"Well," answered the Chevalier, "he might make one or two lights, but hardly more, for it would require too heavy a weight."

"THE SUN of Saturday last printed a comparison of the cost of the electric and gas light. How do the figures tally with your estimate?" I inquired.

Under the circumstances they were a good estimate," the Chevalier replied; "but you compared the cost of producing the electric light with the price charged for gaslight by the gas companies. Now let me make an estimate. We'll say it costs $1 a thousand feet to make the gas. The companies charge their customers double that amount. A burner burns fifty feet a day and one thousand feet in twenty days, allowing ten hours a day for the light. This would be five thousand feet for every one hundred days, and fifteen thousand feet for every three hundred days. The fifteen thousand feet will cost $15. Put in the sixty-five odd days, and we have $18.25, the total yearly cost of producing a gaslight ten hours a day. Now I have figured out the cost of the fuel required to furnish steam power for an electric light of the same degree of intensity, to burn the same period of time daily. If I get six lights per horse power by the use of a new generator, and there is no doubt about it, each light will cost $1.50 a year. To attain this result the best engines must be used. Large engines are the best possibly economy. The figures are based on 3,000 or 4,000 horse power all in one place, with all the modern appliances. I have figured it down. The fuel costs $9 per annum per horse power. For that fuel I get six lights. Bring it down fine and it comes to $1.50 per light. This is for fuel alone, leaving out interest on investment, wear and tear, cost of attendance, oil, and so on. Estimating all these expenses with the poorest engines and the poorest economy, where the expenses are enormous, if I can get six lights per horse power, each light will cost not more than $5.33 per year."

The Chevalier was silent. He went into a musing mood. He looked through the window at a little frog pond for some moments, and then said: "With a one-horse power the Jablochkoff candle gives a light equal to sixty-six gas jets. With the same power I get only six lights, each equal to one gas jet. People look at each other and ask how I can hope to compete with the candle. The answer is easy. The Jablochkoff candle consumes carbon that costs three times more than the power that supplies the electricity. Hence, at the same expense, I could use eighteen of my lights. The candle cannot be subdivided. Its light is so intense that a ground glass globe is used to modify its power. This involves a loss of fifty per cent of light, reducing the value of the sixty-six gas jets to thirty-three. Thus at the same expense we have eighteen of my lamps of one gas jet each to one carbon candle equal to thirty-three gas jets. These eighteen lights, judiciously distributed over the area to be lighted, double their value when compared to a single lamp of eighteen gas jets. What I mean is this: Here is a room with one gas jet. The gas jet is equal to fifteen candles; but fifteen candles distributed around the room would give more light than the gas jet. On the same principle I say that one carbon candle, equal to thirty-three gas jets, would give only about one-half the light that eighteen of my lamps would give. They would actually surpass the Jablochkoff candle in economy when used in lighting up a given area. And all this despite the fact that I can obtain only a total light of six gas jets per horse power to sixty-six by the use of a carbon candle."

The conversation next turned upon the carbon telephone. The Chevalier said that he had sold the right to use the instrument in France for 500,000 francs. Referring to the annual meeting of the American Electrical Society and Mr. Bliss's assertion that he could talk through the Atlantic cable, the Chevalier Edison shook his head. "I think that impossible," he said. He also declared that his receiver for throwing the sound out into a room could not be attached to the Gray or Bell telephones.

15. EDISON'S ELECTRIC LIGHT.[*]

THE GREAT INVENTOR FRANKLY DECLARES IT ENTIRELY SUCCESSFUL.

IT IS IN OPERATION AT MENLO—SEMI-PUBLIC EXHIBITION—
ONLY ONE-THIRD AS COSTLY AS GAS—THE SECRET TO BE
LAID BARE WITHIN THE COMING SIX WEEKS—THE LIGHTS
TO BURN NIGHT AND DAY UNTIL ALL ARE SATISFIED.

Thomas E. (sic) Edison sat in his new office at Menlo Park, N. J., yesterday afternoon looking over specifications for patents. A letter written by a New York correspondent of the *London Times* was laid before him. It says that Edison's electric light is a failure.

"I have seen the letter before," said the great inventor. "It is a mass of misstatements, and is evidently made up in the interests of gas men. The writer is totally at sea."

Mr. Edison then pasted the printed letter upon a large sheet of wrapping paper, and after analyzing and numbering the author's assertions wrote answers to them thus:

1. It is now known that Mr. Edison has failed in his experiment with the electric light.

   "It is not known," Mr. Edison wrote; "quite the contrary."

2. The most that he has ever yet accomplished has been to maintain 400 coiled iron wires in a state of partial incandescence with a

---

[*] From the New York *Sun*, Saturday April 12, 1879, p. 1; clipping in Cummings scrapbook 4, NYPL.

⊰ 134 ⊱

sixteen horse-power steam engine. The object of this experiment was to ascertain the number of coils which could be brought to a red heat in any given circuit. It is upon this experiment that Mr. Edison based his claim that he could maintain 20,000 lights burning from one electrical station with a 600 horse power engine.

"Never made such a statement," wrote Mr. Edison.

3. The conclusion was a fallacious one, as Mr. Edison now knows. Platinum must be heated to 2,700 degrees before it attains the intensity of incandescence which is required for illumination.

"Correct," Mr. Edison wrote. "We heat it every day to 4,000."

4. When the metal is as hot as that it is just on the verge of melting.

"No so," wrote the Chevalier. "See above."

5. To prevent the lamp from melting this inventor has used a regulator, consisting of a bar of metal, through which the current flowed, which, when the current became too strong, expanded and switched off a part of the current, and thus saved the lamp.

"One of 200 methods," the Chevalier wrote.

6. In practice the regulator has failed to perform the service required of it.

"That particular one was not perfectly reliable," wrote the inventor.

7. When the current becomes too strong the platinum burner melts in the twinkling of an eye, and the mischief is done before the regulator can act.

"With the regulator I use," the inventor wrote, "it would be contrary to the law of nature to melt it."

8. The inventor believed that he could overcome this practical difficulty, but he has not succeeded. His lamps have continually melted, and he has been unable to keep them from doing so.

"How does he know this?" was Mr. Edison's written comment.

9. The result is that there is great discouragement at Menlo Park.

"Oh, yes," broke in Private Secretary Griffin, sarcastically. "We're all feeling awful bad."

"On the contrary," wrote Mr. Edison, "people are buying property."

10. There has been another difficulty. Fourteen out of Edison's sixteen applications for a patent at the Washington Patent Office have been rejected.

"This information," Mr. Edison wrote, "could not have been obtained from the Patent Office; besides, nine have been allowed, and eight have not been examined. None have been rejected."

11. This impulsive man took up the electric light last fall as an entirely new subject of experiment, and allowed himself to believe that he saw a way to make the light useful, which others had never thought of, but when he reached the Patent Office he discovered that very nearly every idea which he had embodied in his applications had either been covered by the patents of other inventors or was not patentable at all.

"See above," commented the discoverer.

12. This information is obtained at the Patent Office and is one explanation of the discouragement at Menlo Park.

"Must be a bad leak in the office," Mr. Edison wrote.

13. There is no doubt that the Edison light would be a delightful resource for the illumination of dwellings if it could be depended

upon. It floods a room as though with golden sunlight—pure, brilliant, and mellow. But the inventor has never yet been able to regulate his current so as to keep his lamps burning for any length of time, and has never ventured on a single public exhibition of it.

"Have had a semi-public exhibition for two weeks," wrote Mr. Edison. "Have had 16 lamps burning every evening in the factory and exhibited it to over two hundred persons."

14. The public have never seen so much as one of his lights yet. A favored few, who have been admitted to his laboratory at Menlo Park, have beheld it—a single lamp enclosed in a glass globe, beautiful as the light of a morning star. But he has refused to let any one inspect it closely, and has never allowed the exhibition of it privately to last long. He has never been able to depend upon its durability.

"Just disconnected a lamp this morning," wrote the inventor, "that has burned nineteen consecutive hours and did not show the slightest change."

15. His apparatus is as far from perfection as it ever was, and in fact well-informed electricians in New York do not now believe that Mr. Edison is even on the right line of experiment.

"No electrician in New York has been in my laboratory," Mr. Edison wrote.

16. One of the early friends and associates of Edison was a young man named J. B. Fuller, a Connecticut Yankee, in whom the inventive faculty was active from boyhood. Fuller made a great deal of telegraph machinery and invented many industrial processes, such as modes of utilizing wood pulp in paper making, &c. Ten or fifteen years ago Fuller separated from Edison, because it would no longer do to have them work in company. The inventions of one might easily have been said

to be the suggestions of the other, and so the two men parted never to see each other again.

"I never knew the man," wrote Mr. Edison, "until 1873. I was introduced and spoke to him for a few minutes. I saw him for half an hour twice afterwards."

"There," said the Chevalier Edison, shoving back his chair and slapping the palm of his hand upon the brown wrapping-paper containing his comments. "There is what you want, a full reply to all the statements in that letter."

Mr. Edison says his electric light is a complete success. "If lucky," he says, "we shall have all the houses in Menlo Park lit up in six weeks. The subdivision of the light is perfect. I have had more trouble with the dynamometer for the measurement of the horse power for each generator than I had with the light itself."

The delay in bringing the light before the public has been the want of a perfect generator of electricity. Mr. Edison was satisfied that he could make a machine that would produce double the amount of a Wallace or Gramme machine to the horse power. He has made two machines, and has met with great success. He not only gets more electricity to the horse power, but the current is more applicable to his lamp. One of the generators was tried on Thursday night, and was broken. Mr. Edison says that "it jerked the thing right off the shaft." Batchelor, his assistant, was at work at it yesterday, and both machines will again be in operation in a day or two. So far, the inventor says, he has spent about $30,000 in experiments. He positively asserts that the light is an assured success, and that it can be made three times cheaper than gas. "I mean," he says, "that I can get the same amount of light for three times less money, actual cost to both parties."

The attempt to prevent his obtaining a patent in England failed. He is now getting a new set of patents in all foreign countries. He thinks that inside of two months, if nothing interferes, he will be able to make the whole thing public. He can now get a light equal to thirty candles from a lamp that would at first give a light of only two and a half candles.

"We are safe on our lamps," he says. "The one we propose to use will not melt until it gives out a light equal to three gas jets. We shall not force more than one gas jet from each lamp, using its further capacity as a margin in case of any disarrangement. The latest improvements have rendered the regulator of secondary importance, for it would require extra engines and boilers at the electric stations to produce a force of electricity strong enough to melt the lamps. The only use left for the regulator is to turn the lamp up and down."

Mr. Edison says that he shall burn 500 lights at Menlo Park, keeping them aglow night and day, until the stockholders are satisfied. It has been supposed that the lamps themselves would be very expensive. They are simply a small coil of platinum wire placed in a glass bulb. Mr. Edison says they will cost—bulb, platinum, and all—not more than a dollar and a half apiece. He is making his own bulbs, having picked up the art from a perambulating glass-blower.

A scientific gentleman has tested the colors of the electric light. He found the tints the same as the tints of sunlight. Gas gives no tints.

Sixteen electric lights are now in use in Mr. Edison's machine shop. The electricity has been furnished by a small Gramme machine, but the new Edison generator will be substituted in a day or two. The glass bulbs are no larger than a rusty coat apple. A light equal to that of three gas jets fairly warms them, and that is all. There is scarcely any perceptible heat.

16. WANTED, A PLATINUM MINE.*

THOMAS A. EDISON WILLING TO SPEND $20,000 IN THE SEARCH.

The Present Condition of the Electric Light—A Hundred Men Prospecting for a Mine of Platinum—An Old Trapper's Discovery—Searching the Chaudiere Valley.

On Thursday last the Chevalier Thomas A. Edison stood within his machine shop at Menlo Park, drawing geometrical hieroglyphics and talking scientific jargon with an editor of the *Scientific American*. Such words as "nodules," "globules," "Jablochkoff," "ten Daniels," "dynamometer," "electro-motor," "galvanometer," and "voltaic current," were alarmingly frequent. The soft-moving Brown engine was slashing the electricity out of one of the Chevalier's new generators, and the current was carried to three carbon lights that emitted blinding rays. The great inventor was conducting experiments to illustrate certain abstruse electrical propositions laid down for the benefit of the scientific editor. He bent over one of the shafting wheels, and pressed upon it with an old broomstick. An assistant kept track of the pressure on a Fairbanks scale, and a second assistant filled a book with figures, and announced the result of each experiment. At 3 P.M. the scientific

---

* From the New York *Sun*, Monday July 7, 1879, p. 1; clipping in Cummings scrapbook 4, NYPL.

editor departed, with his head in the clouds and his feet on the earth. The Chevalier then turned his attention to the unscientific reporter of THE SUN. He was about to call his attention to some of his new inventions and discoveries, when the reporter entered a protest.

"Tell me," said he, "the exact condition of your electric light. Do you still use your platinum burner, and are you dead sure that your light is a complete success, and that it will take the place of gas?"

The answer was a positive yes. The Chevalier referred to the specifications of some old patents published in the *Herald* a few weeks ago, saying: "They indicate only a principle, and were the result of my first experiments. They furnish no description of the light as it is now."

"What have you accomplished since those patents were issued?" asked the unscientific reporter.

WHAT HAS BEEN ACCOMPLISHED.

"First," responded the Chevalier. "I have perfected a standard meter for measuring the electricity fed to the burners, the same as a gas meter. It is all right. Second, I have perfected a method of insulating and conveying the wires from the generating stations to the houses of consumers. Third, I have perfected an electric generator. I am satisfied that it cannot be improved. Ninety-four per cent of the horse-power used to run this generator is set free in the form of an electric current. The best machine thus far constructed only frees ninety per cent. To be sure this shows a difference of four per cent only in our favor; but our machine, unlike all other machines, delivers eighty-two per cent of the total power in the wire outside of the machine. It has what is called twice the electro-motor force, or gas pressure, of any machine yet made, with the same resistance of wire and speed. That is why we get nearly double the power out of this generator over any known. I will illustrate more clearly what

I mean. We will say there is a hundred feet of wire wound on my machine and eight hundred feet outside of it. We distribute eight horse-power in an electric current. Eight-ninths of this current reaches the wire outside of the generator and is used for light, and one-ninth is lost in the machine. The other machines may turn the same horse-power, within four or five per cent, into an electric current, but three of the eight horse-power is lost in the generator, and not more than five horse-power reaches the light in the form of electricity. I am supposing that both machines carry a hundred feet of wire inside and eight hundred feet outside. To obtain the result shown by my generator their machine could carry no more than two hundred feet outside. If there was a greater length of wire, they could not transfer the horse-power. The consequence is that one-third of the horse-power is lost in the machine, and only two-thirds is used. The successful generator is the one that delivers the most current outside of the machine, and not the one that transfers the most horse-power into current.

WHERE THE CHEVALIER STANDS.

"The subdivision of the light is perfect, but I am improving the lamp every day," the Chevalier continued. "The latest experiments give me nearly seven gas jets per horse-power, and there are indications that I can increase the number to ten. Just so long as we can see our way to getting more gaslight per horse-power, we shall give no exhibitions. The platinum burner is a settled thing. In all carbon lights not more than forty-four per cent of the horse-power goes into the lamp. We get eighty-two per cent in our lamps. I recognize the impatience of the public over the delay in bringing the light before them, but we must start with a perfect plant. It is a necessity. Suppose we erected our stations and lighted New York city, losing horse-power that might be saved by a perfected lamp or generator. In time the lamps and generators would have to be thrown out and new ones substituted. The company would lose millions of dollars. We are going to perfection even in the supply of metal for burners. I have been bothered to find a dynamometer for measuring the horse-power used to generate the electricity for the light. I made dozens of them before I got one that is absolutely perfect. It measures within a thousandth of a horse-power what goes into a generator and what comes out of it. Such are some of our difficulties. We may be able to spread the whole thing before the public in three or four weeks, and the time may be much longer; but just so long as we can see a chance for improvement, we shall continue our experiments."

Mr. Edison says it has cost him about $13,000 to perfect his generator. He has spent about $8,000 in experiments on his lamp. It cost about $3,000 to discover a new method of insulating his wires. The meter experiments ate up fully $2,000, and the dynamometer $3,000 more. He estimates the total cost of his experiments thus far at $45,000. He says that his patents in foreign countries are all right, and there are big inquiries from Australia.

PLATINUM.

Platinum burners having been definitely settled upon, Mr. Edison is looking for an unlimited supply of the ore. He says he is satisfied that there is any quantity of the metal in the United States, and he can afford to spend $20,000 in finding it. The metal was first discovered in 1741 by Wood, an assayer of the Isle of Jamaica. The ore carries palladium, rhodium, iridium, osmium, ruthenium, and iron. It is found in scales or flattened grains. Sometimes it appears in lumps alloyed with gold, silver, copper, iron, and lead. It appears in alluvial districts in the debris of the earliest volcanic rocks, and permeates the black sand found in auriferous countries.

The Chevalier says he has discovered a sure way of extracting the metal from this black sand. He declares that the ore will yet be found cropping to the surface in a ledge the same as other metals, and believes that it will soon be as cheap as silver.

Russia to-day produces over ten times as much platinum as the rest of the world. The principal mines are near Ekaterinburg, on the Asiatic slope of the Ural Mountains, 180 miles southeast of the city of Perm, on the river Isset. There are mines of copper and iron near by. The platinum is usually found in small scales in veins that run through the mountains. The Demidoff cabinet, however, contains a nugget weighing twenty-one pounds. Russia produces about 3,500 cwt. of platinum annually. The metal is more valuable that gold. For years the supply exceeded the demand. Between 1826 and 1864 Russia used the metal in coinage to the extent of $2,500,000. The value of the coins was eleven and twenty-two roubles.

Borneo produces 500 pounds of the metal yearly, and Ceylon and Brazil are good for similar amounts. The ore is said to be plentiful in the Choco Valley, New Grenada, but the country is so unhealthy that neither whites nor negroes can live within 300 miles of the valley. Choco Valley is near the Atrato River. It was once visited by Humboldt, who brought back a nugget weighing nineteen pounds. This specimen is now in the Royal Cabinet at Berlin. The ore has also been discovered in Santo Domingo, California, British Columbia, and Australia. It has been seen in the sands of the Rhine, in the French Alps, in the county Wicklow, Ireland; in Honduras, in Rutherford County, N. C., and at St. Francis Bianca, Canada.

A CHANCE FOR MINERS.

Mr. Edison began to look for a mine of platinum about the 1st of May. He sent out 2,000 circulars addressed to Postmasters and other public men in mining regions. These circulars read as follows:

FROM THE LABORATORY OF T. A. EDISON,

MENLO PARK, N. J., U.S. A.

DEAR SIR: Would you be so kind as to inform me if the metal platinum occurs in your neighborhood. This metal, as a rule, is found in scales associated with free gold, generally in placers.

If there is any in your vicinity, or if you can gain information from experienced miners as to localities where it can be found, and will forward such information to my address, I will consider it a special favor, as I shall require large quantities in my new system of electric lighting.

An early reply to this circular will be greatly appreciated.

Very truly, THOMAS A. EDISON
MENLO PARK, N. J.

Specimens of platinum and iridosmine sprinkled upon a card were sent with these circulars. The difference in the metals is easily detected with a microscope or magnifying glass. These circulars have also been sent to Sonora. It is said that there is platinum in the Hermosilla mines in that State. Answers to these circular letters are flowing in by scores. Forty-five replies were received on Thursday. The Chevalier has received three specimens of the ore from the Pacific coast. Two of them are rock specimens. The other was taken from the black sand of a sluice box. The miners were throwing it away with the tailings. The latter specimen was very rich. A gentleman who had received a circular called upon the Chevalier recently, and reported that he had found traces of the ore in West Virginia. The following letter from the Pacific slope came on Thursday:

A friend of mine has just handed me your circular letter to answer, thinking that I would be better qualified to answer it than any one he knew, as I have been engaged in mining here for the past twenty years. In reply I would say that platinum is found in most of the placer mines on this coast, but in no place in sufficient quantities to justify mining for it. I believe that I know of a ledge where the metal exists in the same manner as gold quartz. An old hunter and trapper told me of the ledge but a short time ago, and gave me directions for finding it, but I am in so reduced circumstances that I cannot even afford to spend the time necessary to go and prospect it without knowing what I can do with it.

According to my information the ledge is about eight feet wide, and will yield from one-fifth to one-third of an ounce to the pound of rock, taking the lowest estimate, making a yield of 400 ounces to the ton. An ordinary ten-stamp quartz mill would reduce ten tons per day, and allowing twenty-five per cent. for loss in reduction, would leave 3,000 ounces per day net. Now, if such a mine exists, what would it be worth to you?

To this letter Mr. Edison made the following reply:

If you will locate the mine and find it all right, I will furnish capital, put up stamp mill, and give you ten per cent. on every ounce mined. How much money do you require to go and locate the mine and get samples for assay?

Mr. Edison has already received samples and tests from about one hundred miners to whom he has sent his platinum cards. They are now prospecting for the precious metal. It is the most difficult of all ores to reduce, but the great inventor says he can soon discover a method of reduction if he can find a mine.

SEARCHING THE VALLEY OF THE CHAUDIERE.

Not satisfied with these efforts to discover a bed of ore, the Chevalier sent Frank McLaughlin to the gold region of Canada, where he had heard that platinum had been found in black sand.

Mr. McLaughlin left this city on May 26, and returned a fortnight ago. He gives an interesting account of his search. Although the Chaudiere Valley is only about 60 miles from Quebec, the country is little known. It was with difficulty that he could learn the nearest route to the valley. British capitalists had made a determined effort to control the gold mines, and had kept the world in ignorance of their richness. The native miners made a fight for their rights, and they were finally secured under the leadership of Col. Wm. Smart and Louis and John St. Onge. Mr. McLaughlin speaks in glowing terms of the richness of the gold fields. One of the St. Onge claims yields about $4,000 a week. Two brothers Poulin, who went into the country not long ago as poor as church mice, are reported to be worth $100,000 apiece. A poor Catholic clergyman was in want of a little gold for his altar, and one of the brothers offered to give him enough to cover his church. Old prospectors claim to have discovered true fissure veins, and there is a universal cry for capital to develop the mines.

Mr. McLaughlin secured letters from Jules Fancher de St. Maurice of Quebec, and was received by the St. Onges and other French Canadian miners with open arms. He found many sluice boxes containing black sand, almost a sure indication of platinum. He visited mines on the Gilbert River, a branch of the Chaudiere, and found black sand on the St. Onge claim. A shaft had been sunk seventy feet, and a rich bed of pay gravel was struck. A barrel of the gravel carried a pail of black sand. This sand was loaded with gold. There were traces of platinum, but not enough to work up. Traces were also found on the upper River du Loup. On the lower River du Loup McLaughlin saw a ledge that cropped to the surface with a fair sprinkling of the ore. Col. Smart is to prospect for the metal in the branches of the Chaudiere, and Mr. Edison shares the expense.

On his return to Montreal, Dr. Harrington of the Geological

Museum showed Mr. McLaughlin a saucer filled with scales of platinum. They came from the Similkameen River, in British Columbia. A gentleman attached to the geological survey saw the Chinese miners throwing it away with their tailings by the bushel. They were so convinced that it was worthless that he had to pay them to save him a saucer full. As fast as Mr. Edison receives samples of the ore from his prospectors, Mr. McLaughlin will visit the claims and report upon their value.

A SEARCH IN JERSEY.

The Chevalier is terribly in earnest. He wants to find the ore where it will be marketable, and as near as possible to his laboratory in Menlo Park. His thoughts revert from his experiments to the search for platinum. He spent the night of June 26 in scientific experiments. His faithful assistants Bachelor, Griffin, and McLaughlin were with him. Griffin and Bachelor, worn out with work, slid for home about 4 A.M. An hour later McLaughlin tried to follow them. He met the Chevalier at the door of the laboratory. Mr. Edison had an old milk-pan under his arm, and was flourishing a spade with an edge curled like a combing breaker. His eye was bright, despite his night of toil, and he seemed as fresh as a daisy. "Come, Mack," said he, "let's go out prospecting."

There is an abandoned copper mine some distance back of the laboratory. Some wall-eyed Jerseyman had worked it seventy years ago, vainly fancying that he might make a fortune. Some one told the Chevalier that he had seen black sand in the working. His native energy was aroused in an instant. He saddled and "sinched" McLaughlin, and went for the sand. They walked through the tall wet grass in silence. Once in the gully the Chevalier picked out a spot, saying, "Dig there, Mack."

Mack sadly planted a corner of the spade in the soil, and began to heave the earth into the milkpan. Occasionally the Chevalier stooped and picked a leaf or an old clam shell from the shovel. When the pan was filled he waded into a brook, and shook away for dear life. The result was not encouraging. There were no traces of black sand. Back he came, more eager than ever. "Dig away, Mack," he said. "We'll get it yet; we'll get it yet."

So Mack dug, and the Chevalier panned for two hours. The great inventor caught Mack smiling, and was about to say, "Never more be servant of mine," when a gill of black sand attracted his attention. It lay on the bottom of the pan. "We've got it," he joyfully murmured, and McLaughlin joyfully shouldered his shovel.

The black sand was carried to the laboratory and put thorough a careful analysis. The chemist reported that it was as fine a specimen of unmetallic Jersey mud as he had ever seen, and McLaughlin bestowed a parting benediction upon the head of the great inventor, and with wet and weary feet sought his presumable virtuous couch.

17. EDISON'S HORSESHOE LIGHT.*

A CHARRED PAPER BURNER GLOWING FOR ONE HUNDRED HOURS.

THE MENLO PARK INVENTOR'S REPLY TO ELECTRICIAN SAWYER—
PREPARING TO PUBLICLY DEMONSTRATE THAT HE CAN DO WHAT
HE CLAIMS—HOW THE AIR IS DRAWN FROM THE LAMPS.

"There's no unearthly graveyard glare about that," said Mr. Edison, in his laboratory in Menlo Park, last evening, "and if you will notice, there are no sharp shadows."

On a table before him was one of his new horseshoe electric lamps, one of a dozen that were illuminating the low-roofed, dark-walled laboratory. This room, Mr. Edison's office, his shop, his house, and five other houses are now lighted nightly with the new lamps; but this is not the grand illumination promised when all his preparations are made. For this he will fix no date, but hopes to be ready on Monday night next. The new lamp resembles a miniature horseshoe, aglow at a white heat, in a small pear-shaped globe. The loop of fire has a faint orange tinge, and a soft light, not very different from gaslight in color, but purer and without flicker. The reason that the ordinary electric light produces black and sharply-defined shadows is that it proceeds from a very small point. Although it may appear to be as large as a pea, or even as large as a hickory nut, yet a darning needle held between it and

---

* From the New York *Sun*, Tuesday December 23, 1879, p. 1.

the eye will place the eye completely in shadow. The new light, proceeding from a loop over an inch high and three-quarters of an inch wide, shoots rays across the edge of every object from its different parts at varying angles. This causes the shadow to shade gradually off into light.

As one looks with the naked eye at any part of the horseshoe, it appears to be over an eighth of an inch in width, but a pair of blue glasses rob it of its fringe, and it then shrinks to its true dimensions, about a thirty-second of an inch. It then shows like a loop of white-hot iron.

"Suppose, Mr. Edison, I tip that lamp with my cane," said a visitor, "won't that loop of charred bristol board break?"

"I'll show you," replied the inventor.

He then went to a work bench, on which stood a small box of the carbon horseshoes. They had been prepared for lamps, but had proved to be faulty, and were condemned. Laying one of them on a table, he placed a finger on one of the ends of the shoe, and lifted the other up and laid it nearly over on the table in the opposite direction from its fellow. It twisted at the top where the curvature was greatest, and finally broke there, just before the movable end touched the table.

"This is paper," said Mr. Edison. "It isn't merely the remains of bristol board. You may make paper out of various substances, and this is charcoal paper. It has all the texture of the paper left in. All I have done to it is to drive off all the other substances that were in it, by heating it in an oven. The texture remains as it was. The interlaced fabric is left, and it retains its strength."

Then Mr. Edison took one of the lamps and jarred it with his hands without breaking the horseshoe, which could be seen vibrating inside the globe.

One of the lamps hanging from the ceiling had been in use

for five days. It had been burned in all about 100 hours, and the horseshoe seemed as perfect as ever. By Mr. Edison's direction all the other lamps were turned off, and about three horse-power of electricity was allowed to run through a single lamp remaining. It ran up from about eighteen candle power to such a pitch of luminosity that one could read advertisements in THE SUN at the distance of seventy-five feet. The rim of the horseshoe widened until one could not see through it, and the whole resembled in shape an elongated silver dollar. The laboratory is 100 feet long and about thirty feet wide, but all parts of it were illuminated.

Mr. Edison expected that the lamp would be destroyed in one of three ways: either the small platinum wires connected with the horseshoe would be melted, or the glass would crack near where the platinum wires pierced it, or the charred paper would be disintegrated. After vainly waiting until his visitors were tired for the destruction of the lamp, Mr. Edison had the other lamps relighted.

Across a long table in the laboratory ran two wires, three inches apart. If both were touched at the same time, a faint electrical effect was produced on the nerves of the hand. Mr. Edison took several lamps and laid them along between these wires. Each lamp had two wires extending from the bottom. When these wires were attached to the two parallel wires, each lamp in succession burst into luminosity. There were several other lamps near them, fed from the same wires, but their luminosity was not perceptibly diminished. Mr. Edison hopes, in practice, to feed eight lamps with each horse power he uses.

"The secret of this light," the inventor explained, "is that the resistance offered by that little piece of charred paper, about two inches in length, if straightened out, is as great as the resistance of ten miles of telegraph wire. I mean that if the energy of a current of electricity were measured, after it had passed through ten miles of telegraph wire, the result would be the same as if an equal current were measured after it had passed through that little horseshoe. It is this resistance that converts the electricity into heat, and causes the charcoal to glow so brilliantly. All I do is to make the horseshoe a part of the circuit by turning this screw. No matches are needed; no ammonia, naphthaline, or other noxious gases and vapors are given off, and only one-fifteenth of the heat produced by a gas jet is given off."

Mr. Edison laid his hand on the globe of one of his burning lamps and invited his visitors to do the same. The heat was not unbearable, although the glass touched was not an inch from the glowing charcoal.

Mr. Edison said that so much of the air had been removed from the globe by the method he employed to produce a vacuum that its pressure was only one-millionth of that of ordinary air. To exhaust a lamp of air, it is connected with a system of glass tubes. One of these tubes is long and upright. Mercury is forced up through it, driving the air out at the top. The air is not allowed to remain at the top, and when the mercury falls by its own weight, a vacuum is formed above it. This, however, is not perfect enough. By the side of this tube is another long, upright tube. A connection is formed between them so that what air there is in either may circulate through both. The second tube has a small calibre, except where it is enlarged a little to serve as an air chamber, or, as it might be termed, a partial vacuum chamber. If a stream of mercury is made to run down the tube through the chamber, break into drops in the chamber, and then run down the remaining part of the tube in a broken column, between each two portions of the broken column will be carried downward a small portion of air so long as any is left in the chamber. By this process Sprengel was able to produce a nearly perfect vacuum in about twenty-four hours. While Mr. Edison was making use of a Sprengel tube, however, he by accident allowed the mercury to pour through the tube in an uninterrupted stream. To his surprise he found that, although it was not in accordance with Sprengel's theory, the air was exhausted faster than before. In some way, which he does not attempt to explain with certainty, the air was carried out by the continuous stream of mercury more rapidly that by means of the vacant spaces in the interrupted stream. This discovery enabled him to produce such a vacuum as he required in one hour.

The other important processes in the making of a lamp require only skilful glass-blowing and nice manipulation in preparing the carbon horseshoes. It can therefore be made very cheaply. Only a small amount of platinum wire is used. Platinum is the only conductor of electricity Mr. Edison has found that he can pass through glass, and then seal up the opening around it with an air-tight joint. This is because, in the first place, the platinum will adhere to glass in a fused state, and in the second place, because it has very nearly the same amount of expansibility under like degrees of heat.

"Mr. Sawyer says," a visitor remarked to Mr. Edison, "that you cannot run your carbonized paper lamp three hours; that carbonized paper, in practice, in a perfect vacuum will last about twenty minutes."

"Mr. Sawyer couldn't run his lamp three hours," was Mr. Edison's reply, "and I told him so. He doesn't know what I can do with mine as well as you do. You have seen it burning over an hour, and before you go away at 9 o'clock you will have seen it burning over three hours. I think that Mr. Sawyer at the time he wrote his attack on me did not know precisely what he was doing. I shall, in a short time, reply to him and to all other skeptics by giving a public exhibition. I shall light up the ten houses in Menlo Park, and also set up ten electric street lamps. I hope it won't be later than next Monday."

Mr. Edison intends to measure the amount of electricity each family uses in this way. He will draw off from the wire entering the house a certain quantity of the electricity, say one five-hundredth, and let it run through a solution of sulphate of copper. The more electricity that runs through this chemical, the greater the amount of copper that is deposited. This deposit is caught on a small metal plate, to which it strongly adheres. The company's agent, when Mr. Edison makes his monthly visit, is to take this metal plate from the meter and carry it to the office to be weighed. When one plate is removed a clean one will be put in its place.

"Will not your carbon horseshoe consume in the course of time, or waste away?" Mr. Edison was asked.

"I think not," he replied. "I have found that the resistance of the horseshoe, after using for some time, in some cases changes very slightly, but that when it changes it always decreases. This tells me that the horseshoe has grown larger and not smaller, and that none of it has been burned away.

"There is one advantage we shall have over the gas companies," Mr. Edison continued. "We can sell light all night and power all day. The electricity that will run one gas jet will run a sewing machine, and will cost only four cents a day. I have a little motor here with which I have been raising five gallons of water fifty feet high every minute that it was at work, and the electricity used was exactly the amount required to burn one gas jet, that is just one-eighth of a horse power, for I reckon eight gas jets for every horse power."

The Sun.

18. EDISON'S LAMP YET BURNING.*

THE GLOWING HOOP THAT HAS GIVEN LIGHT SINCE FRIDAY LAST.

THE INVENTOR'S CALCULATION THAT THIS IS EQUIVALENT TO ORDINARY FAMILY USE FOR 24 DAYS—HOW THE ELECTRICITY IS OBTAINED, AND HOW IT IS TO BE DISTRIBUTED BY WIRES.

Despatches from London and from Montreal yesterday announced a decline in the prices of gas stock in both cities, and say that the cause of the decline was the receipt of the news that Mr. Edison claims to have perfected his electric light. In this city no public sales of gas stocks have been made, or private ones reported, since the publication of the result of Mr. Edison's tests of his electric lamp in Menlo Park. But probably some sales will be made to-day if there are any offers.

A meeting of the stockholders of the New York Mutual Gas Light Company is called for Jan. 20, 1880, to consider and vote upon the question of reducing the capital stock of the company to $3,000,000. The capital of the company at present is $5,000,000. It also has outstanding $900,000 of its bonds. Several of the directors were asked, last evening, on what grounds the proposed reduction was contemplated. The most communicative of them said that it would be impossible for him, or any of the directors, to make public the reason for the proposed action. The directors,

---

* From the New York *Sun*, Wednesday December 24, 1879, p. 1

who represent nearly three-fourths of the stock, had met and determined upon the reduction, and each had pledged himself not to reveal any of the plan of which the scaling of the capital was a part. To do so might defeat the object in view. As soon as the change was accomplished the reasons for making it would be made public, and the director thought, the public would be satisfied with the action. For the present, not even those stockholders who are not members of the directory are informed of the plans under consideration. He added that a desire to diminish the company's tax account was not a factor in the case, nor would the company curtail its operations. Its capital is now larger than that of any of the other companies, and it will be equal to that of any other company after the reduction. The war between the companies has brought the price of gas down to from $1 to $1.50 per 1,000 feet, according to the quantity consumed by any one establishment. The Mutual Company passed its last dividend.

As to the effect of Mr. Edison's electric light upon the gas industry and the price of gas stocks, the same gentleman declined to venture any predictions. But he expressed the opinion that, however practicable electricity might be for lighting streets and large public edifices, it could never be made applicable to private dwellings. It is not improbable, he said, that a renewal of the discussion of the subject of electric light might produce another such speculation in gas stocks as occurred a year ago. However, the large holders are not the sellers at such a time.

A very large portion of the Mutual stock is held by members of the Vanderbilt family, namely William H. Vanderbilt; his son Cornelius Vanderbilt, who is a director; R. L. Crawford, a brother of the Commodore's widow, and who is supposed to represent her, and Joseph Harker.

The delicate loops of charred Bristol board were yet aglow

yesterday afternoon in Mr. Edison's laboratory in Menlo Park. Some of them have been burning night and day, with the exception of short intervals when the machinery stopped, since Friday morning, so the workmen and other persons in the neighborhood testify.

"My lamps have been burning," Mr. Edison said, "for very nearly 108 hours. Now, the average number of hours per day that a family uses a lamp is four and a half. If you divide 108 by 4 ½ you will get 24. Therefore it has been demonstrated before the eyes of everybody in Menlo Park that one of these lights will last a family twenty-four days. To-day I tested some of those that have been burning longest and found no considerable change. They were tested for 142 ohms, and they were found to be between 141 and 143. I consider that they are quite as perfect as ever. If they have lasted so long without change, I don't see why they should be expected to fail suddenly. Without stopping to make any definite calculations, I know I can make the lamps so cheap that I could afford to supply new lamps to every customer at least as often as once every month and a half."

Efforts are being made to illuminate Menlo Park on Monday evening, but it is thought doubtful whether a new generator that is being constructed will be in readiness by that time. Mr. Edison explained at length his proposed method of laying down wires in the street. Two wires of large size will be laid on each side of the main thoroughfares just under the flags, near the curbs. Neither heat nor cold will affect the currents. Mr. Edison says that he could, if it were not for the increased expense, light up the whole city of New York from one station, but that to do this he would have to have very large copper conducting wires, which would be expensive. The ratio of increase in the expense would be very great as the distance from the station increased. His notion is that the districts should be so small that they can be supplied by means of an engine and boiler and generators that can all be contained in an area of 100 by 25 feet. Buildings of that capacity, he says, may be secured almost everywhere, and at a rent that never would be excessive, because there would always be many such buildings in the neighborhood of a place where it might be thought desirable to have a station.

The wires under the sidewalk are to be enclosed in two parallel pipes with insulating lining, but at every twenty feet there are to be boxes, in the sides of which the pipes will terminate. The wires are to stretch naked through the boxes form side to side. At these boxes the wires may be tapped, either by wires laid along down the side streets or by the small wires leading into stores or houses. The meters, also, may be placed in or near the boxes. It is confidently asserted by the inventor that any number of lights may in practice be obtained from the same circuit, if only enough power and enough generators are employed. The burners at the end of a long circuit will not be as bright as those near the generator, he admits, but he adds that the same is true of gas. He is struck, he says, with the many analogies between the gas and electricity. As in the case of gas, a man will always be on duty at every station to turn on or shut off the current, as may be necessary to keep the lights burning evenly. The difference between the brilliancy of his burners at different points along a circuit will be slight.

"Now, here's a magnet," said Mr. Edison, as he sketched an oblong figure on a sheet of foolscap, "and here's a circuit of wire (drawing a rude circle with his pencil), which may be as long as you please. You want to know just how we get our electricity. Here it is. If I take up a loop of that wire and move it past the end of that magnet a current will be started in the wire. That is the whole mystery. The magnet doesn't make the current. The magnet is a dead piece of matter. Neither does the wire make it. The magnet and the wire loop have enabled me to convert the muscular power of my arm into electricity, and when I move the wire past the magnet there is a certain amount of resistance. It is imperceptible when one wire is used and the magnet is used, but that resistance is a measure of the strength of the current."

Then the inventor explained how a great many wires were wound lengthwise off a reel, or bobbin, which revolves very rapidly between the two ends of an enormous horseshoe magnet. He drew pencil marks across the space between the ends of the horseshoe magnet to represent lines of magnetism. It its passage each wire cuts these lines twice at every revolution of the bobbin, and the bobbin revolves about 700 times a minute. The currents thus started in all these wires are gathered into one current, and its energy measures the power of the machine. Either a permanent steel magnet may be used or a soft iron horseshoe may be kept in a magnetic state by being wound with wire, through which a current of electricity is made to flow. Mr. Edison uses a soft iron horseshoe, and feeds the encircling wires with electricity generated in a second machine. His machine differs from others in this latter respect, but especially in having a very large horseshoe magnet. With it he says he can convert into electricity ninety-four per cent of the power applied to the machine. He finds the percentage in this way. He has devised an easily understood process of weighing (indirectly) his steam or horse power on a Fairbanks scales. Scientists have agreed that a certain cell battery of a certain size will yield a mechanical power that will raise forty-four pounds one foot in one minute. Now, by comparing the electric power obtained from his machine with that obtained from the cell battery of the kind referred to, he is able to calculate the electrical energy he has obtained in terms of horse power. This is the amount of energy that will raise a weight of 33,000 pounds one foot in one minute.

The suggestion recently made that all the power of Niagara

Falls might, with the aid of turbine wheels, generators, and copper wires, be distributed over the State, was mentioned to Mr. Edison. He thought such a thing was possible, but that it would not be accomplished during the next ten years. Buffalo, he conjectured, might be supplied with cheap power from Niagara before that time. The trouble was that the interest on the cost of the necessary copper would make the power thus obtained dearer than steam power.

Mr. Edison does not expect to supply heat over his wires. He does expect to supply power to run all small machines. The heat furnished by his burner is very slight, one-fifteenth of that of gas. If he should have the problem of supplying heat set before him, he says he should study first in the direction of the conversion of power, derived from his electric motor, into heat. The heat thus obtained would be derived from mechanical power, which had been derived from electricity, which had been derived from mechanical power, which, in turn, had been derived from the heat of the original coal. There would thus be a chain of energy of five links.

When the late Prof. Joseph Henry of the Smithsonian Institution in Washington exhibited, in 1831, a machine with electricity as a motor, the belief prevailed in some quarters that steam was to be superseded. Prof. Henry's engine consisted of an oscillating iron beam surrounded by a conductor of insulated copper wire. A current of electricity was sent through this in one direction, which caused one end to be repelled upward and the other attracted downward by two stationary magnets. The downward motion of one end of the beam toward its lowest point brought the conducting wires in contact with the opposite poles of the battery, which produced the reverse motion, and so on continually.

In a subsequent arrangement the velocity of motion was regulated by a fly-wheel, and electro-magnets were substituted for the permanent magnets at first used. After this Prof. Jacobi of St.

Petersburg constructed a large electro-magnetic engine, by which a small boat was propelled several miles an hour. Then Prof. Page of Washington made, at the expense of the Government, a powerful engine of this description. It generated sufficient power to propel a railway car at considerable speed. Modifications of these forms of apparatus have since been made. Like the caloric engine, these machines proved to be failures from a commercial point of view, although successes from a scientific standpoint. It was found that, although the electro-magnetic power could be applied with less loss in the way of effective work than heat by means of the steam engine, yet the cost of the material by which it was generated was so great that it could not be economically used. It was found that one pound of coal in burning develops as much heat as six pounds of zinc; consequently under the same conditions six times as much power was developed from the burning of an equal weight of coal as from the zinc. The power of the steam engine being generated by the burning of coal in air, while that of the electro-magnetic engine was generated by the oxidation or burning of zinc in acid, and since coal and air are the simple products of nature, while zinc and acid require artificial preparation at the expense of power, it was decided that electro-magnetism could not compete with steam.

19. EDISON'S ELECTRIC LIGHT.*

ALL ELECTRICIANS INVITED TO VISIT THE MENLO PARK LABORATORY.

Mr. Edison's Rejoinder to Prof. Morton—His Written Replies to Put Questions—The Light Perfected—Its Cost—Visits from Presidents and Directors of Gas Companies—The Electric Light Still Glowing.

To the Editor of the Sanitary Engineer:

Having a sincere respect for Mr. Edison as an enthusiastic and ingenious investigator, I am sorry to see his name used by writers who evidently are quite ignorant of the subjects about which they treat in a way that will inseparably connect it with discreditable (because false) claims, evidently made in the interest of financial speculators.

No one can more thoroughly appreciate than I do the originality of conception, the indefatigable patience and immense labor which have been involved in the series of experiments of which a sketch has been given in the *New York Herald* of Sunday, the 21st; but when I see the conclusion of these, which every one acquainted with the subject will recognize as a conspicuous failure, trumpeted as a wonderful success, I have only left before me the two

---

* From the New York *Sun*, Saturday December 27, 1879, p. 3; clipping in Cummings scrapbook 4, NYPL.

alternative conclusions that the writer of such matter must either be very ignorant, and the victim of deceit, or a conscious accomplice in what is nothing less than a fraud upon the public.

Such writing as this, in fact, has the melancholy result of placing Mr. Edison and his electric light in the same category with Mr. Keely and his "water motor," Mr. Payne and his "electric engine," Mr. Garey and his "magnetic motor," and others of the same class.

Against this I protest in behalf of true science and for the sake of Mr. Edison himself, who has done and is doing too much really good work to have his record defaced and his name discredited in the interests of any stock company or individual financiers.

HENRY MORTON
STEVENS INSTITUTE OF TECHNOLOGY,
Dec. 22, 1879.

Mr. Thomas A. Edison read the above communication in his laboratory in Menlo Park last evening. His electric lamps were glowing above his head. Turning toward the writer, he said: "Every word that was printed in the article in the *Herald* is literally true, although it was printed without my knowledge. Prof. Morton does not read between the lines. He should investigate first and animadvert afterward. I now authorize you to extend an invitation to him or any other electrician to visit my laboratory and see the light in practical operation. No impediments will be thrown in their way. Everything shall be open and above board. If men however, are willfully blind I cannot undertake to restore their sight."

The answers to the following questions are in Mr. Edison's own handwriting:

"Then you consider your work on the electric light finished?"

"Practically done, though I am still experimenting with a view to reducing its cost."

"What does it cost now?"

"You will have to ask that question of the officers of the company in New York."

"How many lights, each equal to a gas jet, do you get to one horse power?"

"My lights are on a ratio of ten gas jets per horse power per hour."

"What is the power of your engine?"

"Eighty-horse power."

"What does it cost to run your eighty-horse power engine one hour?"

"Seventy-five cents."

"How long do your lights maintain their power without injury?"

"Twenty-three were burning continuously from Friday last to Wednesday, and thirty-three from Wednesday to 10 o'clock on Thursday night. During this time the engine was stopped for an hour to take water. Not a light was injured, and all were regulated at the central station."

"What was the distance of the farthest light that was burning five days?"

"Three lights have been burning that time one-fifth of a mile away."

"Were the twenty-three all connected with one main wire?"

"Yes."

"And more could have been put on the same main wire without increasing the power of the engine or diminishing the light of these twenty-three?"

"Yes, five hundred."

"Have the presidents or directors of any gaslight companies visited your laboratory and seen your light in practical operation?"

"Yes, sir. Mr. Benson of one of the Brooklyn companies and Judge Fisher of the same city have been here. There may have been more. A great number of persons have called. We have refused admission to no one."

"Are any gas men interested in your company?"

"I neither know nor care"

"May I say that you have read this report, and that it is correct?"

"Yes."

When we left Menlo Park the eighty-horse power engine had resumed its work and the chunky little generator was churning out the electricity, scattering a shower of electric sparks over the floor of the machine shop. Two street electric lamps were casting an orange glow upon the snowy meadow in front of Mr. Edison's private office. They materially added to the effulgence of the moon. Twenty odd electric lights were burning in the Edison buildings. It was a whiter light than that in the street. It was more like daylight. The intenseness of the Madison Square Garden and Broadway lights was gone. There was nothing that could annoy the eye. The glow was mild and steady. There was no flicker in the airless globes, and the weakest-eyed sewing woman could have taken her stitches as easily as though she sat by a window at noonday. The engine was running a mass of machinery and working a patent pump, as well as furnishing electricity for the horseshoe burners. This same power can be utilized by those who use the light in their houses. The power that runs the light can run a sewing machine, and the meter will tell the exact power used. The consumer thus pays for the power consumed, whether for the sewing machine or for the light.

20. THE LIGHT IN MENLO PARK.*

MISCHIEF MAKERS IN EDISON'S LABORATORY ON NEW YEAR'S.

———

The Inventor's Temper Disturbed—Lamps that Might be Broken in Gunpowder Without Peril—Some Misstatements Corrected.

So many people visited Menlo Park on New Year's Day that fears were at one time entertained that their weight would break down the second floor of the frame building, which is Edison's laboratory and workshop, where they all congregated. And some of them behaved very badly. Eight electric lamps were stolen. One clumsy fellow got his feet entangled in the rubber hose constituting a part of the delicate apparatus for obtaining a vacuum in the bulbs, and, in blundering his way out, tore down and destroyed the apparatus. A malicious rascal was caught trying to destroy the current and put out the lights by putting a piece of metal across the exposed connecting wires. When he was bounced out, he or other fellows of his kind tore down some of the connecting wires outside and stole the lamps. Even kind, patient, and good-natured Edison was angry before the day ended, and perhaps his ill humor was slightly helped by the knowledge that an inebriated pseudo scientist was in the crowd vilifying him and belittling the result achieved. Yesterday Edison's good humor had returned, but there

———
* From the New York *Sun*, Saturday January 3, 1880, p. 3.

was much greater care and prudence in admitting visitors to the laboratory.

The lights yet glowed steadily all day and night. Some of them have now been burning 218 hours, and are as bright as when first lighted. It is not in any degree true, as has been publicly said, that some of the lamps are burning less brilliantly than when first set up. Some are less brilliant than others, but what they are at the beginning they continue until the end. The explanation of the variance is very simple. Thus far the processes of carbonizing the little paper horseshoes has been necessarily inexact, and, consequently, imperfect. They are made by hand, and practically carbonized by guess. It is intended that each shall have a resistance of 100 ohms, but they may be 95 or 105 ohms without making any material difference. When, however, the resistance rises to 120 or 130 ohms, as it easily may because of imperfect carbonization, a more powerful current is required to produce the desired incandescence than that regulated for the lamps having the lesser quantities of resistance. Out of ten carbon horseshoes made by present processes, six only are good. All this will be remedied in a few days, as soon as some machinery, now being made, is got to work.

After Mr. Edison had been given the foregoing explanation, he was asked about a further affirmation that inasmuch as one-millionth part of atmospheric air remained in the glass bulbs, the destruction of the carbon horseshoes was inevitable. He laughed at it and said: "Mr. Crooks, the London scientist, the discoverer of the radiometer, is the father of high vacuo, and it is from reading his experiments and adapting to our uses the valuable knowledge he gained, that we have been enabled to get this result. Our air pump is a combination of the Geister and Sprengel apparatuses. All the difference is that we have made changes which reduce the time down from forty hours to one hour and a half to obtain a vacuum so nearly perfect that only one-millionth of atmospheric air is left. If we kept on we could get it down to one-half that, but it is not necessary. It is ridiculous to say that in that proportion of air there is enough oxygen to cause the destruction of the carbon.

The process of making the lamps is very interesting. Each is blown in two parts, one a small bulb and tube, the other a large bulb and tube. In the first, two platinum wires each with a minute vise on the end, are hermetically sealed by means of the blowpipe. Then the large one is, in like manner, sealed over the small one, after the carbon horseshoe has been clamped in the vises. Two are connected by tubes about three inches long which join in a single tube, through which the air is extracted from both. Then each is sealed off while it is yet attached to the air-pump and is afterward finished up in the glass house. Notwithstanding the great amount of skilled hand labor now required on these little lamps, Mr. Edison says that so much about them can be made by machinery that he confidently expects to produce them at a cost of not more than 25 cents apiece.

The conducting wires are copper, a little smaller than ordinary iron telegraph wire, and one pair of such wires now run one-fourth of a mile from the generators, support eighteen lamps, and would supply a number more, if desired. Copper is used because it is cheaper for the purpose than iron; but should its price go above thirty cents per pound, iron will be substituted.

A gentleman representing a large cotton mill corporation called upon Mr. Edison yesterday to inquire what it would cost to put in the machinery and furnish 1,000 lights in the mill he represented. The electrician said he was not prepared to give any definite figures, but the amount would not exceed $5,000.

"And what will it cost to operate it?"

"Interest on investment; four per cent for depreciation; cost of power to run dynamo machine. Water power may be said to cost four mills per hour per horse power; say half a cent per hour per horse power by steam. You will require 100 horse power, or perhaps 120. That is something I have not yet fully determined to my own satisfaction. I am making lamps as fast as I can, and intend attaching them all until I find the utmost capacity of the 80-horse-power engine I have here. When that is done I shall know more about it than at present. At present I calculate eight or eight and a half lamps, no more than a quarter of a mile distant, for each horse power. At the same time I shall have a good chance to test the durability of the lamps. I keep them burning steadily day and night. In that way I practically get six days' trial in one, for in ordinary use a lamp is only turned on an average about four hours."

The cotton manufacturer devoted some time to a careful examination of the light, and in subsequent conversation expressed himself delighted with it. One great consideration with him was that its employment would effect a great saving in insurance, for, with the electric lamp, it is simply impossible to set fire to anything, however much one might desire to do so. "Crack one of those glass bulbs in which a light is burning," says Mr. Edison, "and in less than an appreciable space of time the light would have vanished. The carbon would be consumed as quick as the oxygen of the air had reached it. You may break it in powder or nitro-glycerine and its fire will do no harm. I will give five hundred dollars to anybody who will set fire to anything with it."

Even when the little horseshoe was glowing most brightly one could handle the bulb around it without discomfort.

Rumors are in circulation that Mr. Edison has sold out his interest in the light to a stock company. That is not true. He yet holds a considerable quantity of shares in the company which supplied the $60,000 or $70,000 for his experiments, and he has given a liberal interest to his assistants. Drexel, Morgan, & Co. are the principal owners and control England as well as this country. Whether it is true, as rumored, that the Rothschilds have interested themselves in the introduction of the light in Europe, and formed a syndicate with a pledged capital of fifty million dollars for the purpose, he says he does not know and really has no curiosity to know. But he does know that a station for the furnishing of the light for store and household use will very shortly be established in New York—just as soon as his present experiments on the trustworthy capacity of his engine are concluded. The station will supply an area equal to a circle half a mile in diameter; and that much of New York, at least, Mr. Edison says, will have the light in general use within less than six months.

Visitors are warned not to enter the room where the dynamo-electric machine is at work, but they go in all the same. Several valuable watches have already been spoiled by the electricity reaching their works; and one lady on New Year's Day, stooping down to look at something near the machine, was horrified by seeing all her hairpins darting from her head to the generator, beside which she was.

The Sun.

21. EDISON'S LIGHTS NOT OUT.*

RUMORS THAT WERE DISPROVED BY A VISIT TO MENLO PARK.

A Defect, However, Exists in the Glassware of the Lamps that is Exercising the Inventor's Ingenuity— What Mr. Edison Says.

There was published yesterday a statement that matters in the shops of the inventor, Edison, in Menlo Park, are "rather at a stand-still;" that many lamps have suddenly gone out, owing to the breaking of the carbon horseshoes, and that Mr. Edison is now at work solving the mystery of the failure. It was said that the light cracks the lamp glasses and admits air. The article concluded with the announcement that Mr. Edison hopes to overcome these difficulties in time to introduce the light into practical use in New York before next winter. Mr. Edison was shown these statements last night.

"Some of those statements are falsehoods," said he, "and some are quite true. The lamps have not gone out, and the carbon horseshoes are not broken. That some of the lamps have been cracked, and that air has thus gained entrance to the flame, is true. This does not affect my invention. It is purely a mechanical fault—a trouble with the glass. I do not claim to have perfected that part of my lamp. I am still experimenting with it."

* From the New York *Sun*, Friday January 16, 1880, p. 1.

Then Mr. Edison took one of his little glass pear-shaped lamps in his hand and showed how some of them had broken. A few defective ones broke at the top of the dome, a few broke where the inner or wire glass was joined to the big vacuum glass, but the majority broke where the platinum perforates the inner bulk. Edison's electric light is not a hot light. An ordinary gas light is fifteen times as hot. Yet there is sufficient heat emitted by it to expand the parts of the lamps, and it is because the platinum wires expand more than the glass around them that the glass breaks. Mr. Edison's experiments have led him to use many kinds of glass, and the different parts of his lamps have been given many slightly different shapes. All his lamps in the houses, shops, and streets are numbered, and their descriptions are written down. They are examined constantly, and Mr. Edison says he is gradually reaching definite conclusions about the various materials and shapes. He has found out that he can make an almost absolutely perfect lamp by using certain kinds of glass and shaping it by hand, but it would be expensive, and the very first object he strives to attain is cheapness. His lamp must be perfect and cheap also. To be cheap it must be made by machinery. He does not want it to cost more than twenty or twenty-five cents. He makes the big outside bulb of his lamp of the best glass, unannealed and thin enough to be elastic, and therefore stronger for the purposes of a lamp, than if it were thick glass. This cannot be bettered. Only those that were badly made have broken. The trouble is with the inner bulb in its relation to the wires that pierce it. His experiments have shown him that by soldering the platinum wires firmly in places where they pierce this inner bulb with a cement of peculiar white glass that he has obtained from Europe, there is a result of only 22 per cent. breakage. Now, by annealing this glass that composes the inner bulb he believes the breakage will fall to 5 per cent. He will have to build a big annealing furnace, but his lamps, he says, will cost only one cent more than at present. But he says that that will not satisfy him. It compares well with the percentage of broken chimneys in other lamps, but he wants to reduce it to next to nothing, or "one-tenth of 1 per cent," as he expresses it. If he could make the whole inner bulb of this white glass that he uses for solder, his lamps would never break, but, unfortunately, it is a glass that cannot be blown.

"Hang it," he says, in his quick way, "if they would only break right away I could hurry up the experiment, but they don't. Some of them crack in twenty-four hours, some in two weeks; others have been burning four weeks and have not cracked yet. As for their going out, there is this remarkable thing about it—they crack and yet don't go out. I discover that they have cracked, because my microscope shows that the carbon arc, from being a smooth, glossy black, has become rough, has its surface eaten out, and is a deep, soft black, instead of a glossy, hard black. It can always be detected, but there are lamps burning now that began to leak a week ago."

Mr. Edison calls a defect like that a "bug."

"I have had queer experiences with bugs," he said, quite gravely. "Sometimes they'll come right out, and I'll see them and fix them on the spot; but other bugs will hide and puzzle me for a long while. I sat right down and hunted a bug in my telephone for thirteen months, but when I got it the telephone was perfect."

"When all these people go away," he said at a another time, referring to the sightseers that throng his place until the 9 o'clock train takes away the last one, "I'll go to work, and then, you won't think things here are at a standstill."

"What will you do then?" the visitor asked.

"Why," said the inventor, "I'll hunt around with my microscope and—and I'll think."

They say in his shops that there never was such a thinker as he is. By figuring and thinking he does his greatest work. When he wanted to improve on the dynamo machines of the day it is said that he, Mr. Batchelor, and Mr. Upton sat still for three weeks, day and night, figuring and thinking, and thus they arrived at the plan of the machine that is now supplying the sixty lamps in Menlo Park.

Mr. Edison repeated that the only defect that he is combating is in the glassware of his lamp. He says that he has done all that it was said he could not do, in the way of subdividing the electric current, finding a non-destructible burner, inventing a suitable form of lamp for private houses, and producing a cheaper light than gas. It cost $100,000, and it took forty men sixteen months; but now he is only bothered by what no one denied could be done, and the present difficulty may not be remedied in a month, or it may be surmounted in a minute, while he studies it. Meanwhile his employees are preparing plans for the depot he expects soon to establish in New York, with its great engines and its wires to every house in a district of half a square mile. Before that will be done, however, Mr. Edison intends to have increased the number of his lamps to 600 or 700 in Menlo Park. Every house is to be equipped with them, and his shops are to be fitted with as many as possible, because Mr. Edison says that the more lamps he has to study and experiment with the sooner he will remedy the defect that is now his only obstacle.

Mr. Edison has designed a lamp for private houses containing three separate carbon horseshoes. One will be of three candle power, the second of eight, and the third of sixteen. Any or all may be used at one time. He has also designed a new street lamp containing a series of horseshoes in the shape of a pyramid. This, he says, will give as much light as the large lights now in use in this city at about one-half the expense. Mr. Edison exhibited a copy of a written paper last evening, which, he says, has been distributed among the newspaper offices. It is written by a person interested in an electric lamp, and, in trying to prove that Mr. Edison's light is a failure, he makes numerous mistakes, such as averring that carbon is expanded by electricity, whereas it is contracted.

Strangers continue to visit the laboratory. Some have come from as far as Chicago. Representatives of nearly all the large gas companies within a radius of several hundred miles have made a complete study of Mr. Edison's system of lighting. In conversation, these men usually admit the success of the experiments in every respect except economy. They deny that Mr. Edison's light can compete with gas for illumination.

So far as could be learned yesterday there have been no sales of electric stock for some time. The highest bid is $1,200 per share, while none is offered at less than $1,800 and $2,000.

The Sun.

22. MR. EDISON AND HIS CRITICS.*

THE ONE FEATURE THAT HE SAYS IS NOW TO BE PERFECTED.

Intimating that He Laughs Best who Laughs Last—A Schedule of the Hours his Lamps have Burned—A Customer's Experience.

"When Sir Humphry Davy, the famous English chemist, was apprised of the project of forcing carburetted hydrogen gas through a system of pipes for purposes of illumination, he laughed in derision. Nevertheless," continued Mr. Edison, "illumination by coal gas proved to be a great success. My project for the subdivision of the electric light is treated in like manner by all those persons who are profoundly ignorant of the system which I am day by day perfecting. It is a singular fact that persons conversant with the subject, after inspecting my laboratory, are ready to allow that I am all right as far as I have gone. I ask no more."

Mr. Edison insists that the only question now is of the perfect formation of the glass globes of his lamps. This, he says, will soon be brought about. Notwithstanding the occasional unfavorable reports of Mr. Edison's experiments, as published, the company of capitalists who are backing him seem not to lose confidence in the inventor's ability to do all that he has promised.

* From the New York *Sun*, Saturday January 24, 1880.

The proprietor of one of the private dwellings in Menlo Park, that are illumined by the Edison lights, said last evening: "The lights seem almost perfect. They give us absolutely no trouble. When we retire we turn the little screws attached to each and the glow instantly ceases. Often during the day I turn the lights up. So far, nearly three weeks, they have not failed. The so-called scientific persons in New York and others who are continually condemning this plan for lighting as impracticable have a great surprise in store for them. They are sure to see their entire city lighted by these same jets."

A number of new lamps were set up yesterday in the Menlo Park laboratory. To the ordinary observer their intensity does not differ from that of those that have been burning for weeks. Examinations of the carbon contained in these latter, made by means of the photometer and galvanometer show their resistance to be unimpaired. That is to say, the amount of combustion that has taken place cannot be measured.

It is thought in Menlo Park that some of the lights that were stolen during the first week of the public exhibition have fallen into the hands of Western electricians. Should similar lamps be manufactured and used, even without the improved dynamo machines, their introducers will be vigorously prosecuted by the managers of Mr. Edison's Electric Light Company.

A means of preventing the glass tubes containing the incandescent lights from cracking Mr. Edison says he has discovered, and will at once put in practice. At the point where the platinum wires pierce the tubes, a compound is applied having a fixed alkali for a base, and a conchoidal fracture. Several of these, he says, have been burning for eighty hours without even the sign of a crack.

The following are the number of hours that each of the lights now set up at Menlo Park had been glowing up to last evening. It was made by Mr. Edison himself from the book kept by one of his assistants:

597, 465, 465, 515, 508, 503, 517, 487, 487, 532, 190, 200, 200, 430, 390, 400, 367, 370, 370, 373, 332, 332, 332, 332, 332, 417, 326, 336, 283, 286, 294, 294, 300, 300, 290, 283, 301, 367, 215, 381, 238, 194, 193, 194, 194, 136.

This includes not only the lights in the laboratory, but also those glowing publicly in the streets and in the private houses.

Mr. Edison exhibited last evening a series of elaborate drawings which comprise the entire plans for the station that he says he will shortly establish in this city. The machinery is to be placed in a building 25x100 feet. In the cellar is to be five engines of 250-horse power each, making in all 1,250-horse power. The dynamo machines are to be in the second and third stories. Using the small horseshoes instead of those now in use, with the same resistance—100 ohms—Mr. Edison says he finds that he can obtain eleven and one-half instead of eight lights to the horse power. In this manner he believes he can generate 13,750 lights, each having a power of eighteen candles. He says, further, that in the day time, when his lights are not used, he can hire out the power of his engines at great advantage. By means of copper insulated wires he can transmit the power generated, and distribute it among the manufactories within a radius of half a mile, although never more than 50-horse power to any one building. An agent whom he has employed to inquire among the manufacturers in the vicinity of his proposed station has just, he says, reported even more favorably than he anticipated. Mr. Edison went on to say that, although he could only deliver at the end of half a mile 65 per cent of his 1,250-horse power, nevertheless, the profit would be enormous. It stands to reason, he argues, that you can run five 250-horse power engines cheaper than you can run forty-eight 25-horse power engines.

Mr. Edison added that he is now placing his light in a large steamship in course of construction by John Roach at Chester, Pennsylvania. The ship belongs to the Oregon Steam Navigation Company. Lights of three candle power are to be placed in each stateroom. The electricity will be generated by means of a small dynamo machine connected with the donkey engines.

A report from Baltimore says that a day or two ago the best bid for stock of the Edison Electric Light Company was $500 a share, against reported sales three weeks ago at $4,000.

A Paris despatch to the *Telegram* from M. Wilfred Fonvielle, editor of *L'Electricité*, says that the Count de Moncel's attack on Edison in *Le Temps* attracted no attention whatsoever; that de Moncel is not a member of the Institute, as is supposed, and that in reality he is what is known as a *membre libre*, without pay and without vote, enjoying only secondary privileges. Moncel is said to be editor of the paper connected with Würdemann's lamp, which is specially intended for lighting large spaces, and which will be set aside by Edison's discoveries.

Mr. Jean Baptiste Dumas, the well-known scientist, said to a Paris correspondent of the *Telegram*: "I own that on first reading of Edison's discoveries, I was incredulous, but I now see no real impossibility in the thing, I do not doubt that the carbonized horseshoes may be held by the platinum clamps, for I have obtained in my own laboratory carbonized Bristol board of a certain cohesiveness. I could only render it incandescent, however, for the space of a few centimetres [the centimetre is about thirty-nine one-hundredths of an inch] by using the Bunsen pile of twenty elements. I am led to conclude that Edison must employ considerable force, and it is difficult for me to understand how he can long render incandescent a carbonized substance with such conductivity as the drawings represent."

M. Níandet Breguet, nephew to the great electrical manufacturer, said: "I found nothing very new or astonishing in the discovery by Mr. Edison, except the manner of fabricating the carbon horseshoes. I find the electrical magnetic machines employed by him excessive, and that consequently there must be a loss of magnetic force in the generation of the electricity.

M. Fontaine, President of the *Syndicat d'Electricité*, of which all Paris electricians are members, and editor of the *Revue Industrielle* said: "I have read everything published on the subject. My conviction is that by the employment of the carbonized horseshoe, Mr. Edison has made an important advance in the application of the electric light. I believe that this result has been obtained by a certain dexterity of manipulation, of which Mr. Edison has preserved the secret, as, for instance, the judicious employment of the electric current during the process of carbonization."

"This is no Keely motor business out here, and people are at last waking up to what I'm doing," said Mr. Edison, when he spoke of the favorable criticisms of the French electricians. "Dumas, too, is an authority. He is mistaken in reference to the 'high electrical resistance in the incandescent substance,' however. It is exactly the quality we want, for it is the enormous resistance of the kind of carbon that I use that allows of the practical subdivision of the electric light."

23. EDISON AND HIS VISITORS.*

THE THREE LADIES TO WHOM HE WAS INTRODUCED THE OTHER DAY.

A LITTLE TRICK THAT DID NOT WORK—THE INVENTOR'S IMPRESSION OF HIMSELF AS VIEWED BY LADIES—STEALING THE QUEER LITTLE LAMPS—THEIR PERMANENT FORM NOT YET DEVISED.

To any one who has seen the electric lights that are displayed in various parts of the city, Mr. Edison's light must at first seem a very feeble illuminator. The passengers by the Pennsylvania Railroad invariably scramble to that side of the cars from which Edison's lights on the depot steps at Menlo Park are visible; but when the lights are seen they are either pronounced a failure or those who see them declare that they cannot be the lights they were looking for, but gas jets. In the darkness of a country roadside the light they shed seems at a distance to produce only a glimmer.

When a passenger quits the train and, mounting the stairs, obtains a close view of the lamps and perceives the horseshoe shape of the flame he is assured of two certainties—that they are Edison's lights, and that their illuminating power is at least that of an ordinary gas jet. There is no brilliant, fierce flame such as is produced by what New Yorkers know as the electric light. In its place is a tiny hoop of white fire, bent at the top and fastened at

---

* From the New York *Sun*, Sunday February 8, 1880.

the ends, and this seem to glow like white hot metal rather than to blaze. Once on the brow of the land the visitors see these lamps fixed to ordinary lampposts placed at short distances one from another and following the lines of the country roads. They look like gas lamps, except that gas is yellow and they are white. They are not encumbered with any mechanical appliances as the old electric lights are, and take up less room than a flame of gas. There is nothing about them but a pear-shaped lamp chimney containing two wires and the horseshoe of fire. This is enclosed in a great glass globe such as our city lamps have. A plate of wire gauze covers the bottoms of these globes, because when the lamp was first exhibited many were stolen. The visitors to Edison's shops saw him and his employees lifting the lamps in and out of their sockets, and they made a tour of the country roads and lifted out twenty-three lamps to carry home and exhibit. One of the stolen lamps is on exhibition in Rahway, another in a Broadway window, and the whereabouts of several others is known. Since the wire gauze protectors have been used to cover the openings in the lamp globe, one lamp has been stolen. The man who took it got it by smashing the great globe with a cobble stone.

Mr. Edison grieves over the loss of these lamps because they were of various kinds, and he was studying their behavior from day to day in order to determine which shape or material or combination of devices was the best. As THE SUN has already stated, Edison keeps a book in which he enters daily whatever has been noticed about each of the sixty lamps that are kept burning. He says that these lamps differ one from another. The difference, however, is in such small points that none but a practised eye can detect it. It is when the lamp is seen burning indoors that the first feeling of disappointment is overcome. There is one of these lamps in the sitting room of a house near the depot. It can be seen from the road through the great bow window of the room. The room seems to be brilliantly illuminated by but one lamp, although only a few feet away a street lamp of exactly the same power is feebly struggling with the darkness, just as a gas lamp would in the same situation. The truth is, as Mr. Edison always explains to his guests, the Edison lamp was designed for use in houses, and it is meant to have the same candle power as a gas jet. Nearly every house in the village contains one or more of these lamps. Whoever remains over night there dines by this light and finds it in his bedroom. The Menlo Park people are all enthusiastic partisans and praise the lamp as if they were all stockholders in the new company.

"Those other fellows," said Mr. Edison, in speaking of his fellow electricians, "began at the big end of the horn with a tremendously powerful light. They tried to reduce their light down to a serviceable power. They are stuck fast somewhere in the horn. I began with a light of two candle power, and I have raised it to the power of an ordinary house light." Mr. Edison was standing in his office with a well-worn high hat on his head, a handkerchief around his neck and knotted under one ear, and with his hands in his pockets. Two or three persons were talking to him and twenty others were staring at him. He had a pencil and a piece of paper in his pocket, and he used these in calculations continually. He solves arithmetical problems rapidly and accurately, and it is said that he leaves to conjecture nothing that figures will demonstrate. He is said to have astonished some gas men who found that he had at his fingers' ends the details of their business. He assured them that he had not based his estimates upon any reports by the close monopolies of this country, but upon the official Parisian reports covering very many years. During one hour the other day he made many calculations—once to find out the average breakage per day among a million lamps if the average life of a lamp was one year;

another sum was worked in order to discover the percentage of breakage thus far among his sixty lamps, and a third concerned future operations in this city. Mr. Edison had been talking about the near prospect of the introduction of his light in New York. A draughtsman was busy with the plans for the interior of the great depot from which the illuminating power is to be distributed over a square half mile of city houses. Mr. Edison had been saying that a recent law permitted the laying of wires under the streets, and that he would take advantage of this.

"But how will you secure customers for your light?" he was asked.

"I'll trust to human nature for that," said he. "If I can furnish an article as good as gas at a cheaper rate, I am sure it will be purchased. I shall say to the people who live in the district I propose to begin with: 'Here is a light that gives out one-fifteenth the heat of gas, that gives a steady light, that is cheap, that can set fire to nothing, that does away with matches, that can be managed by a child, that can be turned upside down, that is ornamental, and that can be converted into a propelling power—to run a sewing machine, for instance—when it is not needed as a light.' Suppose a person wishes to illuminate his house, I can furnish 200 or 300 extra lights, and put them wherever the mistress of the house decides. Afterward they can all be taken down with but little trouble, and but little cost for the whole operation."

Edison's shops are almost as public as the City Hall; consequently, in the daytime and early evening a never-ending line of visitors march through the different buildings. It is at 9 o'clock, after the last train has carried off the last visitor, that Edison and his assistants do their hard work. It was while Mr. Edison and the writer were walking and talking together one evening last week that the inventor was informed of the desire of several ladies to "see" him.

"Tell them," said he, good naturedly, "that the man they saw blowing glass in the little building to the right was me. Tell 'em it was me incognito. They will be just as well satisfied." Whether this deception was attempted or not, the women, three in number, suddenly came across Edison in one of the shops, and one of their male companions introduced them to him before he could escape. He looked perplexed, then smiled, bowed afterward, and finally plunged past the women through an open door.

"Well, they saw you," said the writer.

"Yes, and they must have been astonished to see such a stupid fellow," Edison replied.

Mr. Edison will not increase the number of his lights until he has decided upon the most economical and efficient description of lamp. There are eighty-five lights now up. Among these are some of the old lamps, some of the new, and a lamp with a carbonized hemp horseshoe, instead of cardboard. Again, in some the horseshoe carbons are attached to the wires by platinum clamps, while in others this connection is made by clamps of lignum vitæ, and again of plumbage, made up almost entirely of carbon, and containing little iron. The eighty-horse power engine is at work night and day. Besides the lights, which Mr. Edison says are generated at the rate of from 8 to 10 per horse power, this engine runs all the belting and shafting for the big workshop back of the laboratory.

The Sun.

24. BERNHARDT AT MENLO PARK.*

EDISON MEETS THE MOST INQUISITIVE OF HIS INTERVIEWERS.

INVESTIGATING THE SECRETS OF ELECTRICAL MACHINES WITH INTENSE INTEREST—GREAT ADMIRATION OF THE GREAT INVENTOR'S WORKS.

One of a party of six that were waiting for a belated train in the dimly lighted depot at Menlo Park late last evening was a slight young woman wrapped in a profusion of velvet and furs, who, half reclining upon one of the benches, uttered a variety of strange sounds in an unnatural tone. The other five members of the party took the deepest interest in the performance, which was occasionally interspersed with a chattering of French and a merry laugh. The young woman was Sarah Bernhardt, and the strange sounds she was making were in imitation of those of the phonograph which the Wizard of Menlo Park had just been operating for her benefit.

Having accepted an invitation to visit Menlo Park last night, she did not allow the bad weather to prevent her going. She was promptly on time for the train, which left the Pennsylvania depot in Jersey City at 6:35 P.M. During the hour's ride she asked innumerable questions about the young electrician, his home, and his works. Her curiosity was abundantly gratified by the

* In New York *Sun* on Monday November 29, 1880, p. 1.

ingenuity of her companions. When their accounts of marvellous things became too extravagant, she would cast glances of inquiry at her fatherly manager, Mr. Henry C. Jarrett, whose significant head shakes prevented her credulity from being imposed upon. At Menlo Park the party climbed a long flight of icy steps, at the top of which a small conveyance awaited the ladies. The gentlemen tramped across the vacant logs in the snow and mud to the library. After they were reunited on the library steps they were escorted up stairs. Mlle. Bernhardt, eager as a child, followed closely after Mr. Jarrett. Through a small labyrinth of bookcases they came at last upon Mr. Edison, seated at his desk, poring over books and papers. He had ceased to expect his guests. He rose, looking embarrassed. Bernhardt stepped forward to meet him, her face a picture of delight. They shook hands cordially, and then the actress turned imploringly to the friend nearest her. She had been told that Mr. Edison could not speak French. As she looked she rattled off a long sentence in French. Her friend, acting as interpreter, told Mr. Edison that Mlle. Bernhardt was delighted to meet him, and admired his wonderful genius. Mr. Edison bowed. Then he excused himself in order to get some of his engines in motion, so that he could display some of the results of his labors. Meanwhile Mlle. Bernhardt studied intently everything about her.

In a few moments the electrician returned and led his visitors down stairs and out into the rain and mud to his laboratory. It has the appearance of a workshop, and was dimly lighted by a few flaming, smoking gas jets. Edison brought his guests before a large cupboard and opened its doors. The actress clapped her hands and delivered a volley of exclamations. Upon the shelves of the closet were ten or a dozen small electric lamps. Each gave a light of about the brilliancy of a gas jet. Not one of them flickered or changed its intensity. The aggregate light was as mild, pleasant, and devoid of color as the sunlight. The Bernhardt lavished more words on the surprise. Edison stood aghast at her volubility. With his smoothly-shaven face giving him a boyish appearance, his lips parted in a half-embarrassed smile, his brow was knit evidently in a vain endeavor to get an inkling of her meaning, and with a small black turban cap cocked up on his head he looked every bit the curiosity that the actress evidently considered him. She looked at the lights and chattered like a magpie, except that her tones were musical, and then she would pause and look at the inventor. After Edison became a little accustomed to her volubility he attempted, through an interpreter, to explain the construction and operation of the lamps. But before the half was told her she knew it all. Then she fired a volley of questions. She wanted to know when she could have the light in a theatre, when in her home, when it would be introduced in Paris, whether she could operate the machinery herself, &c., &c. She revelled in the light because it did not make her look green; she eulogized its purity and merits innumerable that she discovered in it. She held up different portions of her dress to see how the various colors looked in the light. From the lamp Edison went to some of the electric machines he has invented. One for measuring the intensity of a light particularly interested the actress. In examining them and everything else she discovered the mechanism and action before the explanations could be completed. She tripped over lumber and went through more mud and rain to the machine shop. Every object interested her. She took pieces of machinery in her light gloves, and soiled them still more by touching greasy lathes and dirty tools. When she reached the room containing the machines that generate the electricity she seemed to realize that the time had come to settle down to the business of investigating. The monotonous whir seemed to inspire her. She threw aside her wraps, that she might more easily follow the inventor through all the narrow, intricate passages between the machines. Her costly dress dragged against oily journals and caught the drippings from swiftly revolving shafts. Her one thought was to see. An assistant attached some iron wire to one of the machines. It burned to a white heat, then fell in pieces on the floor. Mlle. Bernhardt picked them up and wound them into a little coil for a souvenir. When she passed out of doors again several miles of lamps along the railroad track had been lighted. She was taken into another building and shown that the turning of a small disc a few inches in circumference extinguished all those lights as well as many in the buildings. A reverse movement relit them. She was filled with admiration. Having exhausted the lights, she asked for the telephone, and was taken to one where she heard the inevitable, "Hello! hello!" business prior to the singing and whistling of a young man in Mr. Edison's house, a quarter of a mile away.

The performance decided her to have telephones as well as electric lights just as soon as she returns home. Limited time prevented as long an examination of the wonders of Menlo Park as she desired. Her last treat was an exhibition of the phonograph. Edison talked into it, sang into it, whistled into it. Then he and an assistant sang into it at the same time. They attempted to sing different parts; the result was discord, but Edison ground out the hymn—it was "John Brown's Body"—and to show his appreciation of his own musical efforts he went over the same strip of foil a second time, shouting "Shut up!" "Police!" "Get Out!" &c. When he ground out the hymn the second time with the derogatory interruptions Bernhardt's delight seemed complete. Finally she recovered from her laughing sufficient to deliver a passage from Phèdre, and another from Hernani into the machine. Edison had preceded her with a rapid rendition of "Yankee Doodle." The two were ground out together. The contrast provoked another burst of merriment.

Leaving the shop and library the actress was driven to Edison's home, where she was introduced to Mrs. Edison, whom she congratulated upon the possession of so famous a husband. From the house she returned to the depot, where she amused her friends during the half hour they were obliged to wait for the train by her observations upon what she had seen.

The Sun.

EPILOGUE: "IT SHINES FOR ALL"*

The end of 1880 marked the beginning of Thomas Edison's transition from inventor to business person intent on monetarizing light bulbs and electricity while making them available to the public. Amos Cummings, as he had been during the previous three years, was there to chronicle Edison's newest activities in the pages of the *Sun*, beginning with an article published January 7, 1881.

That article, the last of the seventeen listed in Cummings's scrapbook, is headlined "Edison's Electric Light," with "An Exhibition—The Difficulties of Introducing It in the City" as the sub-heading. It recounts a January 6 visit to Menlo Park by some of New York City's best-known financial entrepreneurs, including investors in the Edison Electric Light Company founded by Edison in October 1878 to raise funds for his laboratory (see Chapter 9).

According to the *Sun* article, among the "score of capitalists and stockholders" who arrived at Edison's laboratory on the 5:30 evening train were J. Pierpont Morgan and J. Hood Wright of the New York City financial firm Drexel, Morgan & Co. The company was a financial supporter of the Meno Park lab. Egisto Fabbri of the firm Fabbri and Chauncy, whose offices were also in New York City, was there as well. He later would suggest a plan to Edison to install electricity in several South American countries. Some of the "capitalists" also were thinking about expanding into Europe.

Edison was ready with his show. He turned a dial to the right,

---

* "It shines for all" was the *Sun's* motto.

and "In an instant 800 electric lights were aglow within a radius of a mile…. The laboratory, the machine shops, and many private dwellings were…illuminated…. The glossy beavers [top hats], gray moustaches, gold-rimmed spectacles, and ivory-handled umbrellas of the capitalists were brought into bold relief." Edison had done it!

Edison, ever a showman, turned the dial back to the left and the lights were extinguished. Then he turned them back on a second time, again flooding "the snowy fields with light." While the lights continued to illuminate the scene, he turned on "two [electric] sewing machines, a pump, and a blower" using the same electric source.

Edison announced to Cummings, "The light is now perfected and I am now awaiting the action of the company." The investors— Morgan, Hood, and the others—were sold. The visitors left to catch the 6:30 train back to New York City. There were fortunes to be made.

By the end of 1881, J. Hood Wright had hired Edison to install a generator and electricity in his mansion, located on 176th Street in Manhattan in what was then known as the suburbs. It was the first private New York City residence to be electrified. The next year J. Pierpont Morgan would follow suit, electrifying his new mansion at Madison Avenue and 36th Street in the upscale Murray Hill district of Manhattan.

Two months after Hood and Morgan visited Menlo Park, Amos Cummings caught up with Thomas Edison at the new offices of the recently-founded (December 17, 1880) Edison Illuminating Company at 65 Broadway in New York City's financial district. Cummings's interview was published in the *Sun* on March 15, 1881. The article "Edison Shoving His Light" (number sixteen in Cummings's scrapbook) describes the office building on Broadway as "aristocratic," while Edison is said to be "clean-shaven. A shining beaver was perched on his head, and he was puffing a fragrant cigar.

He was arrayed in holiday attire, presenting a marked contrast from his usual appearance in Menlo Park."

Edison confirmed his new status, "I have left the laboratory... and am now a man of business. The electric light is perfected in all its branches, and I am spending all my time and energies to its introduction to the public." The Edison Illuminating Company soon founded similar companies elsewhere in the United States.

In 1891 one such company, the Illuminating Company in Detroit, would hire Henry Ford as an engineer. Ford, who was experimenting with gasoline engines and predecessors of the automobile, would leave the company in 1899 to start his own automobile company. He and Edison would remain friends.

To say Edison's electric company was a success is an understatement. It was bought out by Consolidated Gas in 1901; the owners of New York City's gas utility company could see the future. Eventually Consolidated Gas would become Consolidated Edison. Today, Con Edison (also known as ConEd) provides electricity to several million people in New York City.

And the Edison Electric Light Company? It merged with another company in 1892 and became General Electric. The next year Edison sold his shares of GE for $1.5 million (about $430 million today). No wonder Thomas Edison's light bulb has come to symbolize a good idea, a very good idea.

www.ingramcontent.com/pod-product-compliance
Lightning Source LLC
Chambersburg PA
CBHW071424170526
45165CB00001B/383